AGRICULTURAL ISSUES AND POLICIES

ADVANCES IN HYDROPONICS RESEARCH

AGRICULTURAL ISSUES AND POLICIES

Additional books in this series can be found on Nova's website under the Series tab.

Additional e-books in this series can be found on Nova's website under the e-book tab.

AGRICULTURAL ISSUES AND POLICIES

ADVANCES IN HYDROPONICS RESEARCH

DEVIN J. WEBSTER
EDITOR

Copyright © 2017 by Nova Science Publishers, Inc.

All rights reserved. No part of this book may be reproduced, stored in a retrieval system or transmitted in any form or by any means: electronic, electrostatic, magnetic, tape, mechanical photocopying, recording or otherwise without the written permission of the Publisher.

We have partnered with Copyright Clearance Center to make it easy for you to obtain permissions to reuse content from this publication. Simply navigate to this publication's page on Nova's website and locate the "Get Permission" button below the title description. This button is linked directly to the title's permission page on copyright.com. Alternatively, you can visit copyright.com and search by title, ISBN, or ISSN.

For further questions about using the service on copyright.com, please contact:
Copyright Clearance Center
Phone: +1-(978) 750-8400   Fax: +1-(978) 750-4470   E-mail: info@copyright.com

NOTICE TO THE READER

The Publisher has taken reasonable care in the preparation of this book, but makes no expressed or implied warranty of any kind and assumes no responsibility for any errors or omissions. No liability is assumed for incidental or consequential damages in connection with or arising out of information contained in this book. The Publisher shall not be liable for any special, consequential, or exemplary damages resulting, in whole or in part, from the readers' use of, or reliance upon, this material. Any parts of this book based on government reports are so indicated and copyright is claimed for those parts to the extent applicable to compilations of such works.

Independent verification should be sought for any data, advice or recommendations contained in this book. In addition, no responsibility is assumed by the publisher for any injury and/or damage to persons or property arising from any methods, products, instructions, ideas or otherwise contained in this publication.

This publication is designed to provide accurate and authoritative information with regard to the subject matter covered herein. It is sold with the clear understanding that the Publisher is not engaged in rendering legal or any other professional services. If legal or any other expert assistance is required, the services of a competent person should be sought. FROM A DECLARATION OF PARTICIPANTS JOINTLY ADOPTED BY A COMMITTEE OF THE AMERICAN BAR ASSOCIATION AND A COMMITTEE OF PUBLISHERS.

Additional color graphics may be available in the e-book version of this book.

Library of Congress Cataloging-in-Publication Data

ISBN: 978-1-53612-131-5

Published by Nova Science Publishers, Inc. † New York

CONTENTS

Preface vii

Chapter 1 Irrigation Management Techniques Used in
Soilless Cultivation 1
*G. Nikolaou, D. Neocleous, N. Katsoulas
and C. Kittas*

Chapter 2 Nitrate-Ammonium Ratios and
Silicon Used in Hydroponics 33
*Renato de Mello Prado, Rafael Ferreira Barreto
and Guilherme Felisberto*

Chapter 3 Interactive Effects of Copper and Lead on
Metal Uptake and Antioxidative Metabolism of
Centella asiatica under Experimental
Hydroponic Systems 57
*Chee Kong Yap, Ghim Hock Ong,
Wan Hee Cheng, Rosimah Nulit, Ali Karami and
Salman Abdo Al-Shami*

Chapter 4 Phytoaccumulation of Heavy Metals from Water
Using Floating Plants by Hydroponic Culture 73
Anil Kumar Giri and Prakash Chandra Mishra

Chapter 5	Development of Nutritive Solutions for Hydroponics Using Wastewater *André Luís Lopes da Silva,* *Micheli Angelica Horbach,* *Gilvano Ebling Brondani and* *Carlos Ricardo Soccol*	91
Chapter 6	Hydroponics for Feasibility Test of Biodegraded Fishery Waste/Wastewater as Biofertilizer *Joong Kyun Kim, Hyun Yi Jung, Ja Young Cho* *and Geon Lee*	101
Chapter 7	Human Urine Associated with Cassava Wastewater as a Nutritive Solution with Potential to Agricultural Use *Narcísio Cabral de Araújo,* *Abílio José Procópio Queiroz, Rui de Oliveira,* *Monica de Amorim Coura, Josué da Silva Buriti* *and Andygley Fernandes Mota*	137
Chapter 8	Application of Hydroponic Culture for the Cultivation of Edible Cacti *Takanori Horibe*	153
Expert Commentary: Plant Factories and Edible Cacti *Takanori Horibe*		**173**
Index		**183**

PREFACE

This book begins by describing the irrigation management techniques that are currently used for scheduling irrigation in a hydroponic culture, presented in Chapter One. Chapter Two covers research information about the nitrogen forms used in hydroponics, their relationships and peculiarities, as well as information on the use of silicon, its available sources and role in mitigating the toxic effects of ammonium. In Chapter Three, Centella asiatica was tested for the effects of Pb exposure on Cu accumulation and antioxidant activities which includes the activities of ascorbate peroxidase (APX), catalase (CAT), superoxide dismutase (SOD), and guaiacol peroxidase (GPX). Chapter Four discusses the phytoextraction potential of the free floating aquatic plant for heavy metals from aqueous solution. Chapter Five analyzes the use of wastewaters to formulate nutritive solutions represents a rational alternative to wastewaters disposal and adds value to what is currently considered a waste product. Chapter Six provides a brief review of hydroponics used in fishery waste reuse. Chapter Seven discusses the results of physicochemical characterization of Human urine, Cassava Wastewater and alternative nutrient solution prepared through these effluents as an alternative for its use in agroecological systems of agricultural cultivation. Chapter Eight describes recent progress and findings on the hydroponic culture of edible Opuntia. Finally, the book concludes with an expert commentary that argues that plant factories are powerful tools to cultivate and improve the quality of edible cacti.

Chapter 1 - This chapter describes the irrigation management techniques that are currently used for scheduling irrigation in a hydroponic culture. Precise irrigation should involve the determination of the timing and the quantity of each irrigation event which may be estimated based on the climate of the greenhouse, monitoring of the substrate, or evaluated different plant indicators of water stress. Depending on the crop growth stage, irrigation could also be used as a tool for stimulating vegetative or productive growth, or affecting the rate of drainage emissions and thus the quantity of water and nutrient outflow from the greenhouse to the environment.

For many years, irrigation was applied to crops at fixed time intervals and quantities (i.e., time clock scheduling) and, more recently by estimating the quantity of solar energy corresponded to the irrigation dose consumed by the transpiration. However, it has been well documented that none of these two methods are sufficiently accurate to satisfy the crop irrigation needed when used as a solo criterion for irrigation. In principle, a time lag between water supply and transpiration often occurs in the case of time clock scheduling, while irrigation based on solar radiation is not taking into account other climatic factors which affects transpiration, such as the vapor pressure deficit.

Therefore, irrigation scheduling should be based on more complex evapotranspiration models, which correlate to greenhouse climatic and plant data. As the climate equipment used in modern greenhouses allows for the computerized prediction of transpiration rates on a real time basis, the water used by the plant can be estimated accurately in a short interval rate. Under these circumstances a variety of evapotranspiration models which were originally developed by the Penman-Monteith equation which can be effectively used in soilless greenhouse production. However, the model equation coefficients must firstly be calibrated for the specific crop under the prevailing greenhouse environmental conditions. Other than direct measurement of plant transpiration, lysimeters may be the only way of calibrating evapotranspiration models, as it has been used to great success over the past few decades for the estimation of transpiration for field and greenhouse cultivations as well.

Alternatively optimal approaches to irrigation control may include plant stress sensing as opposed to climate or substrate irrigation based approaches. Physiological real-time monitoring for assessing the dynamics of plant water status, seems to be of great value for tuning irrigation and developing crop water stress indices.

Chapter 2 - Excess nitrate in the nutrient solution of some hydroponic crop species has attracted the attention of regulatory agencies because this form of nitrogen can be often toxic to humans, especially children. Inclusion of ammonium in hydroponic nutrient solutions can promote significant increase in productivity and thus, help meet the growing demand for safer foods. However, several factors may alter the availability of ammonium to plants, which may lead to phytotoxicity. However, silicon can mitigate abiotic stresses such as ammonium phytotoxicity, and can be used to prevent possible damage to plants from this form of nitrogen. Therefore, the objective of this review was to gather research information about the nitrogen forms used in hydroponics, their relationships and peculiarities, as well as information on the use of silicon, its available sources and role in mitigating the toxic effects of ammonium.

Chapter 3 - *Centella asiatica* is a medicinal plant that has widely been used for therapeutic purposes in this region. In this study, this medicinal plant were tested upon for the effects of Pb exposure on Cu accumulation and antioxidant activities which includes the activities of ascorbate peroxidase (APX),catalase (CAT), superoxide dismutase (SOD),and guaiacol peroxidase (GPX). The plants were grown in a hydroponic system under laboratory conditions. The results showed positive antagonistic effects of a second metal (Pb) on the uptake of the other single metal (Cu) by *C. asiatica*, which suggested a competition between the uptake of these metals by the plant. All four antioxidative enzymes in the *C. asiatica* showed significant increment when exposed to different metal treatments (showing the enzyme activities increased in the order of Cu>Pb). In comparison between single and combinational effect of the metals, the degree of enzyme activities in combination metals fluctuates. The present findings indicated irregular patterns of Cu accumulation whether at lower or higher levels of single or a combination of both Cu and Pb exposures. Although the four antioxidative enzymes (CAT, GPX, APX and SOD) are potential biomarkers of toxicities of Cu and Pbexposures in *C. asiatica*, further studies are still needed.

Chapter 4 - Advanced technologies and technical progress water contamination by various pollutants is one of the most significant environmental problems more widespread in the future. The present aimed to develop the phytoextraction potential of the free floating aquatic plant for heavy metals from aqueous solution. The accumulation, relative growth and bio-concentration factor of metal ions at different concentrations of chromium solution significantly increased ($P<0.05$) with the passage of time.Plants treated with 4.0 mg/L of chromium (VI) accumulated the highest concentration of metal in roots (1320mg/kg, dry weight) and shoots (260mg/kg, dry weight); after 15 days. To ascertain the mechanism of the process the plant biomasswas characterized by SEM-EDX and FTIR techniques. Microwave-assisted extraction efficiency is investigated by comparison of the results with after wet digestion. Chromatograms are obtained for chromium species in plant shoot biomass by using HPLC-ICP-MS. For extraction of chromium ions from plant materials using two extractant solution, 95.14% extracted by 0.02 M ethylene diaminetetra-acetic acid (EDTA), and 90.24% extracted by a HCl at 80°C duration 25 minutes. Phytoextraction technique is environmental friendly, cost-effective, aesthetically pleasing, technologically feasible, long-term applicability, and ecological aspect.

Chapter 5 - Hydroponics is a technique of growing plants without soil, in water containing dissolved nutrients. Nowadays, the hydroponics has several applications, such as: (1) Aid in the acclimatization of micropropagated plants, (2) Treatment of wastewaters, (3) Method to produce foods in places of the earth or outside that cannot support crops in the soil, (4) Sustaining the growth of ministumps or microstumps to produce cuttings for plant propagation, (5) Allowing to collect roots without kill the plant to obtain active principles present in the roots in the case of medicinal plants and (6) allowing crop production in larger amount and quality. However, the practice of hydroponics is expensive and one alternative to reduce the production costs is to develop nutritive solutions from by-products. Several wastewaters that can be used to develop these nutritive solutions are the vinasse, cassava wastewater, corn steep liquor and domestic effluent. Mineral nutrients must be determined to establish the chemical supplementation in some cases. Sometimes, decantation and filtration of wastewaters must be carried out to avoid toxicity to the crops. Adequate dilution of the wastewater is important to avoid salt stress (e.g., only 10% vinasse was used to develop a vinasse nutritive solution for lettuce, rocket and watercress). Crop productivity using wastewater solutions is similar to the commercial solutions used for hydroponics. The use of wastewaters to formulate nutritive solutions represents a rational alternative to wastewaters disposal and adds value to what is currently considered a waste product.

Chapter 6 - This chapter provides a brief review of hydroponics used in fishery waste reuse. First, the advantages and disadvantages of hydroponics and the application of lab-scale hydroponics to the fertilizing ability test of biodegraded fishery waste are discussed. Finally, an advanced building-type hydroponics model is proposed with discussion of its characteristics and possible problems in operation.

Recently, the paradigm for waste policy has shifted to highly efficient social-based resource recycling, moving toward zero-emission waste management. Fishery waste is biodegradable, and the resultant culture broth is reused as biofertilizer without further wastewater treatment. The success of this process is mainly dependent on the fertilizing ability of the culture broth. To assess the feasibility of the culture broth as biofertilizer, hydroponics is a useful tool. Case studies of lab-scale hydroponics using the culture broth support hydroponics. Based on these results, an advanced building-type hydroponics model is proposed and discussed for practical use. Eventually, the practical use of the proposed hydroponic model is expected to result in development of modern hydroponics technologies.

Chapter 7 - Agricultural use of wastewater containing essential nutrients for plants is an alternative to minimize environmental pollution and overexploitation of mineral fertilizers used in agriculture. In this context, this work aims to present and discuss results of physicochemical characterization of Human urine, Cassava Wastewater and alternative nutrient solution prepared through these effluents as an alternative for its use in agroecological systems of agricultural cultivation. The treatment and physicochemical characterization of the wastewater was carried out at the Federal University of Campina Grande. Following standardized methodology was characterized the hidrogenionic potential (pH), electrical conductivity (CE), chemical demand of oxygen (CDO), chloride (Cl⁻), total phosphorus (Pt), orthophosphate (P-PO_4^{-3}), ammoniacal nitrogen (P-PO_4^{-3}), N-NH^{4+}), total nitrogen Kjeldhal (TNK), potassium (K), sodium (Na) and thermotolerant coliforms. After analyzing and comparing the results it was concluded that Human urine and Cassava Wastewater presents significant quantities of nutrients and can be used in family farming as an alternative source of fertilizers; Solutions of Human urine associated with the Cassava Wastewater present potential to be used in an agroecological culture system, aiming at the recycling of nutrients as a sustainable alternative to minimize the environmental impacts caused by the inappropriate practices of wastewater discharge into the environment.

Chapter 8 - Hydroponic culture is a method of growing plants using nutrient solution (water and fertilizer) with or without the use of an artificial medium. The lack of soil means absence of weeds or soil-borne disease, thus making precise fertilizer management possible. Thus, hydroponic culture conveys numerous advantages for edible *Opuntia* production, which is conventionally produced through soil or pot culture. To date, the authors have investigated the effects of hydroponic culture involving the deep flow technique on the growth of edible *Opuntia* and showed that can be grown effectively using hydroponics compared with the commercially practiced pot culture. Thus, hydroponics may be an ideal method of edible *Opuntia* cultivation for farmers and horticultural hobbyists based in cities, who typically practice soil and pot culture. In this review, the authors describe recent progress and findings on the hydroponic culture of edible *Opuntia*.

Expert Commentary - Plant factories are horticulture greenhouses or automated facilities where vegetables and other crops can be produced throughout the year by controlling environmental conditions such as light, temperature, humidity, CO_2, and nutrient availability. Although food safety and constant supply of food crops to the market is the main advantages of plant factories, many new techniques have been recently invented to increase the productivity and nutritional content of vegetables and to produce high-quality and value-added products. Hydroponic culture is used as a basic cultivation system in plant factories. The authors have investigated the effects of hydroponic culture [deep flow technique (DFT)] and cultivation conditions on the growth of edible *Opuntia,* which is conventionally produced through soil or pot culture. The authors' study showed that the growth of cladodes was greatly affected by growth conditions in the plant factory. Edible *Opuntia* has many features suitable for the production in plant factories, which require large running costs. For instance, it grows rapidly and can be vegetatively propagated through stems (in the short term until harvest). Additionally, *Opuntia* can be planted on cultivation panels at high density (effective use of available space) under relatively low light intensity using artificial light. Therefore, the authors propose that plant factories are powerful tools to cultivate and improve the quality of edible cacti.

In: Advances in Hydroponics Research ISBN: 978-1-53612-131-5
Editor: Devin J. Webster © 2017 Nova Science Publishers, Inc.

Chapter 1

IRRIGATION MANAGEMENT TECHNIQUES USED IN SOILLESS CULTIVATION

G. Nikolaou[1,], D. Neocleous[2], N. Katsoulas[1] and C. Kittas[1]*

[1]Department of Agriculture Crop Production and Rural Environment, School of Agricultural Sciences, University of Thessaly, Volos, Greece
[2]Department of Natural Resources and Environment, Agricultural Research Institute, Nicosia, Cyprus

ABSTRACT

This chapter describes the irrigation management techniques that are currently used for scheduling irrigation in a hydroponic culture. Precise irrigation should involve the determination of the timing and the quantity of each irrigation event which may be estimated based on the climate of the greenhouse, monitoring of the substrate, or evaluated different plant indicators of water stress. Depending on the crop growth stage, irrigation could also be used as a tool for stimulating vegetative or productive growth, or affecting the rate of drainage emissions and thus the quantity of water and nutrient outflow from the greenhouse to the environment.

---

[*] Corresponding author: gnicolaounic@gmail.com.

For many years, irrigation was applied to crops at fixed time intervals and quantities (i.e., time clock scheduling) and, more recently by estimating the quantity of solar energy corresponded to the irrigation dose consumed by the transpiration. However, it has been well documented that none of these two methods are sufficiently accurate to satisfy the crop irrigation needed when used as a solo criterion for irrigation. In principle, a time lag between water supply and transpiration often occurs in the case of time clock scheduling, while irrigation based on solar radiation is not taking into account other climatic factors which affects transpiration, such as the vapor pressure deficit.

Therefore, irrigation scheduling should be based on more complex evapotranspiration models, which correlate to greenhouse climatic and plant data. As the climate equipment used in modern greenhouses allows for the computerized prediction of transpiration rates on a real time basis, the water used by the plant can be estimated accurately in a short interval rate. Under these circumstances a variety of evapotranspiration models which were originally developed by the Penman-Monteith equation which can be effectively used in soilless greenhouse production. However, the model equation coefficients must firstly be calibrated for the specific crop under the prevailing greenhouse environmental conditions. Other than direct measurement of plant transpiration, lysimeters may be the only way of calibrating evapotranspiration models, as it has been used to great success over the past few decades for the estimation of transpiration for field and greenhouse cultivations as well.

Alternatively optimal approaches to irrigation control may include plant stress sensing as opposed to climate or substrate irrigation based approaches. Physiological real-time monitoring for assessing the dynamics of plant water status, seems to be of great value for tuning irrigation and developing crop water stress indices.

1. INTRODUCTION

Soilless culture is considered one of the main components of sustainable protected horticulture (Massa et al., 2010). In fact, the adoption of protected soilless culture cultivation with the implementation of technical practices such as the integrated plant protection and climate control systems, as cited by Meric et al. (2011); positively affected crop growth and yield, increased the water use efficiency and minimized the harmful effects of agro chemicals to the environment. Without doubt, farmers have improved their irrigation strategies and minimized nutrient outflow (e.g., nitrates and phosphates) from greenhouses to the environment. This takes into account the scarcity of water resources and local legislation regarding groundwater aquifers which may preexist in multiple countries (Morille et al., 2013). However, the concept of precise irrigation, indicates that there is considerable scope for improvement of irrigation management practices, especially in open soilless systems, where drainage rates of 30-40% of the nutrient solution applied would be recommended.

Apart from time clock scheduling, which is probably among the most common used method for monitoring irrigation; irrigation amounts are generally determined according to crop water use, so the accurate estimation of crop evapotranspiration is a key parameter in the water management for greenhouse cultivation (Liu et al., 2008; Abdel-Razzak et al., 2016). Despite the Class A evaporation pan being one of the most widely used climatic measurements to determine evaporation, and has been used to great success over the last decades for estimating evapotranspiration in the field. Nowadays evapotranspiration inside highly automated greenhouses is best estimated through transpiration models on a real time basis. Measuring plant water consumption by utilizing complex radiation or energy balance models, weather data correlation (i.e., temperature, humidity, radiation) and plant related characteristics (i.e., leaf area index, leaf temperature, leaf aerodynamic and stomata resistances) with water use, have proved helpful in irrigation control in soilless production systems (Morille et al., 2013).

Since the 1990s, several studies have been conducted, focusing on different greenhouse microclimates and how they are affected by different cooling systems (e.g., fan and evaporative pad, forced ventilation, fog) in order to evaluate the best model for predicting real time transpiration along with different irrigation strategies (e.g., Baille et al., 1994; Katsoulas et al., 2006; Tsirogiannis et al., 2010; Villarreal-Guerrero et al., 2012). The majority of these studies use a variety of evapotranspiration models originally developed by the Penman-Monteith equation (Schröder and Lieth, 2002). For example, the simplified form of the Penman-Monteith equation, allows for the prediction of the actual evapotranspiration rate by overcoming difficult to measure variables of the plants, for example leaf resistances, with a precise level of accuracy using commonly measured weather data (Baille et al., 1994). Thus, many researchers have successfully used the simplified Penman-Monteith method for estimating transpiration for several ornamental and horticultural greenhouse crops (Baille et al., 1994; Kittas et al., 1999; Pollet et al., 2000; Montero et al., 2001; Rouphael et al., 2008). However, the accuracy of the model is crop specific and highly depends on the microclimate of the greenhouse (Yang et al., 1990; De Graaf and Esmeijer, 1998; Medrano et al., 2005, Fazlil Ilahi, 2009). Hence we can directly measure transpiration through the use of lysimeters on plants actual water consumption, with the aim of model calibration as weight changes in mass, and could be one of the most accurately measuring the quantity of water over shorter intervals (Meijer et al., 1985; Beeson, 2011; Schmidt et al., 2013).

Even though in the field of modeling transpiration, a significant amount of research has been conducted, a better understanding of the interactions between the microclimate and the physical conditions of the plants, could be obtained as far as the prediction of irrigation; it's based on plants' actual responses to changes in a plant water status. Therefore, in any case of monitoring irrigation, a range of crop physiological parameters such as growth, photosynthesis, transpiration, which fit within the "speaking plant" approach, may be automatically monitored by a variety of instruments, either remotely or through physical contact (Ehret et al., 2001; Helmer et al., 2005; Steppe et al., 2008; Fulcher et al., 2012; Zhao et al., 2016). In such cases, it is of great value for the system to be in a position to continuously and automatically monitor groups of plants, rather than individual plants which may not provide representative data related to the plant water status (Helmer et al., 2005). It seems that by monitoring plants, we get a better understanding of the plants' physiological status, allowing adjustments on irrigation application techniques. A possible future perspective for scheduling irrigation could therefore include the use of remote non destructive plant reflectance indices for the detection of plant water stress.

With respect to the above mention subject, the objective of the present chapter is to do so, to indicate the most common applications for irrigation scheduling in hydroponic cultures.

2. TIME CLOCK SCHEDULING

Time clock scheduling is considered to be the simplest and the cheapest technique of irrigation adopted by growers (Oki and Lieth, 2008). The frequency of irrigation is controlled according to a fixed program, which may be easily reprogrammed depending on different plant growth stages. It usually commences one hour after the sunrise, and finishes one hour before sunset, with irrigation intervals approximately every hour. Even though time clock scheduling is more reliable than other irrigation techniques, many times, results in a high percentage of drainage and suboptimal root conditions which negatively affect plant growth, as it not always in agreement with the crop transpiration rate values. Incrocci et al. (2014) concluded that there was considerably less water use and nutrient emission depleting into the environment when the irrigation scheduling was based either on the substrate water status or on calculated crop evapotranspiration, as opposed to time clock scheduling.

Lizarraga et al. (2003) who evaluated time clock scheduling, with irrigation intervals of an hour, indicated that tomatoes growing in perlite during the afternoon hours were facing water stress conditions, even though the drainage amount was almost 46% of the daily irrigation water applied. On the other hand, as the frequency of irrigation increased from two times, to 10 times (at one-hour intervals) and to 18 times (at half-hour intervals), it positively affected the yield of lettuce (Silber et al., 2003). However, in any case of time clock scheduling, supplement monitoring of the volume of the drainage amount (i.e., 20-30% of the daily nutrient irrigation solution) ensured that there will be no salt build up in the substrate.

3. SCHEDULING IRRIGATION BY SOLAR INTEGRATORS

As radiation affects transpiration by approximately 70%, the frequency of irrigation could be controlled through water supply in proportion to the accumulated radiation, which is calculated from the sum of photons that reach a defined area over a certain unit of time (Savvas et al., 2007; Shin et al., 2014). However, this method does not take into account the effect of vapor pressure deficit (VPD) on transpiration (Lizzaraga et al., 2003). Consequently, it is necessary to consider VPD for transpiration calculation when using ventilation, heating, cooling or dehumidification systems; even though, according to Kittas (1990) and Katsoulas & Kittas (2011) heating systems have a small influence on crop water consumption which is practically insignificant. Additionally, the daily inaccuracy in water, needs to be estimated by up to 1 mm, owing to the successive approximations in solar radiation (in the case where there is no real time, available data of solar radiation), the value of the transmissions and the value of the crop coefficient (Kittas, 1990). In addition to that, transpiration rates not always increase proportionally to the accumulated radiation levels. It depends on the region and climate, as well as crop growth stages and development and light intensity; which varies with the time of day and season (Shin et al., 2014). Therefore an accurate measurement of light intensity is required for an accurate calculation of radiation accumulation which reflects on to the transpiration rate.

During daylight and especially under high intensity light conditions, the irrigation intervals increase as the transpiration rates increased, resulting in a midday irrigation interval time of less than 30 minutes multiple times a day. In contrast, at low light conditions the start time of the first irrigation even may be delayed by more than two hours after sunrise as a result of low transpiration rates. Hence in order to have a complete irrigation approach when using solar integrator is to combine it with other irrigation techniques such as time clock scheduling, in order to avoid the dryness of the substrate, especially during the hours of darkness and to account for the transpiration of plants. There are several detailed reports on night time plant transpiration, such as Medrano et al. (2005), which reported cucumber night transpiration values between 120 and 200 gm^{-2}d^{-2}. According to the same author, at high radiation levels, the nighttime transpiration was a minor part of the total daily transpiration, but at low radiation levels; nighttime transpiration was as high as 20% of the total daily transpiration. Similarly, Carmassi et al. (2013), reported that estimated gerberas night transpiration values, accounted for 12% in spring and 8% in the autumn of a 24 hour transpiration rate values/cycle.

However, even with those limitations, solar radiation is still widely used as a method for irrigation scheduling and several authors have written reports on the accumulation radiation values for starting irrigation. According to Schröder and Lieth (2002), common solar radiation values, measured inside greenhouses, they range between 0.4-0.6 MJ m^{-2} and 1.4-1.8 MJ m^{-2} in closed and open hydroponic systems respectively; resulting in approximately 30% and 15% of the nutrient applied, depleting into the environment. Additionally, Katsoulas et al. (2006) concluded that when working with different irrigation frequencies in soilless rose crop, the accumulated solar radiation outside the greenhouse reached 1.6 MJ m^{-2}, in favor of rose cut flower, rather than increasing the irrigation interval set points to 3.2 MJ m^{-2}.

For estimating the irrigation frequency set point values, the following equations were used:

$$E = \frac{T_r}{(1-D)} \quad (1)$$

$$Tr = \zeta R_{GO} \quad (2)$$

$$\zeta = K_c \tau \frac{\alpha}{\lambda} \quad (3)$$

where 'D' is the drainage rate, 'Tr' is the the crop transpiration in kg m^{-2}, 'Kc' is the crop coefficient, 't' is the greenhouse cover transmission to solar radiation, 'a' is the evaporation coefficient which represents the part of the energy of incoming solar radiation that is transformed to latent heat through transpiration; and finally 'λ' is the latent heat of vapourisation of water in kJ kg^{-1}.

However, the equations presented, indicate that the transpiration rate is affected by the crop coefficient (Kc) which is not constant; therefore prompting the need to change radiation value levels during different crop phases (e.g., stage of growth, harvesting, removal of old and damaged leaves). Similarly, for tomato growing in perlite substrate, with a plant density of 1.6 pl. m^{-2}, Lizarraga et al. (2003) proposed indoor values for starting an irrigation even of 0.81 MJ m^{-2} (0.4 mm). Additionally Meric et al. (2011), indicated that when the plant density increased to 3.48 pl. m^{-2} the accumulated radiation values of 2 MJ m^{-2} (0.8 mm), resulted in higher yield parameters compared to 1 MJ m^{-2} (0.4 mm) and 4 MJ m^{-2} (1.6 mm).

4. SUBSTRATE MONITORING

In practice, if a producer knows how different kinds of substrates react to different volumes of irrigation applied, he could schedule or implement irrigation, based on a specific substrate water content and electrical conductivity (Lee, 2009).

4.1. Volumetric Water Content and Matric Potential

An approach to irrigation scheduling entails the use of root zone sensors to monitor substrate's moisture status and to replenish the water in a growing medium to a preset level, based on substrates measurements of either water potential or volumetric water content (Pardossi et al., 2009). In spite of the availability of different types of sensors such as tensiometers, volumetric water content sensors, neutron probes, time domain reflectometry probes, their use for irrigation scheduling in hydroponic production is rare and limited, mainly due to their high costs, unsuitable size, and unreliable measurements, as cited in Nemali et al. (2007). Even so, sensors that estimate substrate volumetric water content such as the

TDR-type sensors, have a propensity to be more reliable as opposed to sensors measuring water availability through the matric potential (e.g., tensiometers, psychrometers) (Jones & Tardieu, 1998; Murray et al., 2004). This is particularly true as the difference between the water content at the substrate suctions between 1 to 5 kPa, which represents the "easily available water" (EAW); is not taking into account the accessibility of the "available water", to meet the evaporative demand or the potential rate of transpiration (Wallach, 2008). Thus, due to saturated and unsaturated conditions which occur at the substrates even during irrigation events; they affect the hydraulic conductivity. Therefore, unsaturated hydraulic conductivity, is a better choice of to be use for controlling irrigation in containers filled with porous substrates as it indicates better values than the actual availability of water to the roots; being a characteristic function of the medium and are highly sensitive to moisture variation (Raviv et al., 1999).

In spite of this limitation a significant amount of research has been conducted, as the amount of volumetric water content in the substrate affects the plant's growth rate and production (Fulcher et al., 2012; Fields et al., 2016). As cited in Schröder and Lieth, (2002) in order to increase root growth and gain faster yield from the spring crops' cucumber cultivation, a slightly substrate water stress after transplanting must be performed through keeping the substrate moisture level at 60%. Nighttime irrigation should apply, only if the moisture of the substrate has fallen below 8-10% of the moisture of the previous morning. Additionally, Saha et al. (2008) reported that substrate volumetric water contents' set point values of tomato growth in rockwool 70% or 500 gr. of weight loss, for substantially better root growth and earlier yield, instead of increasing the substrate volumetric water content to 80%. According to the same author, when slap volumetric water is used as a sole criterion for the irrigation decision, it failed to control the slap electrical conductivity to the desired levels, so it is important to combine it with other sensors when making a decision on what irrigation system to utilise.

4.2. Electrical Conductivity

The availability of substrate moisture sensors, that are also capable of measuring permittivity, temperature and substrate bulk electrical conductivity (ECB), have great potential in soilless irrigation management; even though different equation models are required to be used relating the ECB to pore water electrical conductivity (ECP). Monitoring of the substrate electrical conductivity could be used as a tool to stimulate crop development, improving fruit quality, indicating whether or not there is a need for substrate flushing or increasing the water use efficiency by keeping the root electrical conductivity closed to maximum permissible crop limits and automatically having control of saline water irrigation schemes (Valdés et al., 2014).

It is well known that substrate, low EC conditions could lead to crop vegetative development, while higher EC values combined with a slight amount of water stress, stimulates generative development. Accordingly, in order to stimulate generative development and create a large volume of roots, on a bright day, the substrate EC on tomatoes growing in rockwool, could be risen up to 5 to 6 dS m^{-1} and at the same time decreased the substrate VWC to 50-55%. Moreover, under low radiation levels the substrate electrical conductivity could be further increased up to 8 dS m^{-1} (Lee, 2009). Additional, as cited by Nebauer et al. (2013), when the electrical conductivity of the nutrient solution is increased to 4.0 dS m^{-1}, an increase in the maximum photosynthesis rate of tomatoes by 14-15% was recorded. Further increasing to 8.75 dS m^{-1}, indirectly affects plant growth, as it does not induce changes in the photosynthesis rate of the leaf, but diminished total leaf area, thereby reducing whole-plant photosynthesis.

However, as it has been consistently investigated by several authors, prolonged periods of high substrate electrical conductivity resulted in a decrease of water inflow to the fruits, which resulted in lower mean fruit weight (e.g., Savvas & Lenz, 2000; Magan et al., 2008).

In the case of ornamental plants, increasing the electrical conductivity even for short periods negatively affected production. According to Baas et al. (1995) the maximum salinity without yield reduction for gerbera grown in perlite substrate, is 1,5-2.8 dS m^{-1} and for each unit increased above the threshold of electrical conductivity, the flower production decreased by about 10%. Similarly Carmassi et al. (2013), shows that an increase of 1.4 dS m^{-1} in average electrical conductivity of the recirculation nutrient solution for gerbera plants grown in rockwool, resulted in a reduction of flower production of approximately 19% in different salinity levels during spring while no significant effect on all measured quantities were observed in autumn.

It is therefore recommended that salinity effects should be taken into account according to the crop, in order to prevent over application of saline water and hence preventing yield loss. On a final note, the fact that the efficiency of water use degreased, while water salinity increased (Mahjoor et al., 2016).

5. CROP EVAPOTRANSPIRATION

Evapotranspiration is affected by multiple factors; environmental (e.g., air temperature, radiation, humidity and wind speed), plant related characteristics (e.g., growth phase, leaf area index) and the type of the growing substrate and container size (Bacci et al., 2008). The amount and the timing of irrigation, may even be estimated indirectly, based on evapotranspiration models, or directly by measuring differences in crop mass at short time intervals; accounting for transpiration; especially, in high tech greenhouse constructions, where data acquisition on evapotranspiration could be implemented on a real time basis, allowing adjustments on irrigation dose and timing.

5.1. Pan Based Irrigation

Class A evaporation pans, are one of the most widely used systems for climatic measurements, in order to determine evaporation. It has been used for the estimating amount of evapotranspiration in irrigation in a daily basis or even longer periods; for open field and greenhouse cultivations. The simplicity and high degree of adaptation on farm's levels, which are characteristic to the evaporation pan, seems to be the primary reason for the pan expansion between producers; even though within greenhouses smaller pans or atmometers are preferable, as they occupy less space (Imtiyaz, 2000; Folegatti, 2001; Liu et al., 2008).

The estimating of crop evapotranspiration with the use of a class A evaporation pan, can be calculated according to Allen et al. (1998), by using the following equation:

$$ET_C = E_p \, K_p \, K_c \qquad (4)$$

where 'ET_c' is the maximum daily crop evapotranspiration measured in mm, 'E_p' is the daily evaporation from class A Pan in mm, 'K_p' is the pan coefficient and 'K_c' is the crop coefficient.

The difficulty of obtaining accurate field measurements with the use of pan A in herbaceous instead of orchards plants, is due to the fact that; the crop coefficient (Kc value) is constantly changing throughout the growth, pruning and harvesting within a few days period. Orgaz et al. (2005), working in soil cultivations, indicated that the Kc values of vertical support crops like melon, green bean and sweet pepper were between 1.3 and 1.4 and they were higher than those estimated of the same crops cultivated outside the greenhouse. In contrast the Kc values of melon and watermelon (not supported) were between 1 and 1.1, similar to those measured for the same crops under field conditions. Research by Rouphael & Colla (2005) on soilless cultivation of zucchini squash (*Cucurbita pepo* L.), indicated that the crop coefficient (Kc) during the spring-summer period ranged from 0.10 to 1.15, while for the summer-fall season the Kc ranged from 0.12 to 0.80. Furthermore, according to Tüzel et al. (2006), Kc values in an open system, were found to be slightly higher than the closed system with maximum values for tomato grown in a perlite of 1.6.

Despite the extensive use of a Class A evaporation pan, in soilless culture there is limited information on their application. Sezen et al. (2010) working in soilless tomato crop in ash and peat (1:1) substrate, concluded that the optimum amount of irrigation was 1.50 times of the daily Class A pan evaporation value, for twice a day irrigation, under glasshouse conditions in the Mediteranean Region.

Even though pan evaporation can be used in different types of greenhouses, with the simplest one being highly automated; nowadays evapotranspiration inside highly automated greenhouses is more accurately estimated through using transpiration models on a real time basis, as the environment of the greenhouse could change drastically within minutes.

5.2. Model Based Irrigation

Irrigation water amounts are generally determined according to the crop's water use, so the accurately estimating of crop evapotranspiration is a key parameter in water management for greenhouse cultivation, which allows for implementation of transpiration models for automated irrigation control (Liu et al., 2008; Carmassi et al. 2013). In greenhouses, crop evapotranspiration is influenced by the energy balance of the whole system which strongly depends on the characteristics of the greenhouse and on the climate control equipment (Fazlil Ilahi, 2009). Air temperature and vapor pressure deficit are parameters affecting the thermal and hydrological negative feedback effects existing in a greenhouse. Therefore the main factors affecting greenhouse crop transpiration are solar radiation, vapor pressure deficit, the canopy and aerodynamic conductances (Katsoulas and Kittas, 2011).

Different models have been used for irrigation control in protected cultivation (e.g., greenhouses, screenhouses) based on the originally developed evapotranspiration equation of Penman-Monteith. Yet the P-M equation model was primarily developed for open field conditions; based on a big-leaf theory and perfectly mixed-tank approach, that assumes homogeneity of both the thermodynamic conditions within the canopy and the air above the plants (Morille et al., 2013). For hydroponic crops, which were mainly cultivated in high technology greenhouses, transpiration rates could be estimated with high accuracies through Stanghellini and the

Fynns' models. However, such models require difficult to measure variables such as; the aerodynamic and canopy resistances of crops surfaces (Fernández et al., 2010).

In order to overcome those limitations several authors' (Kittas et al. 1999; Pollet et al., 2000; Montero et al., 2001; Rouphael et al., 2004) used the following simplified form of the Penman-Monteith equation, according to Baille et al. (1994).

Table 1. A (dimensionless) and B (Kg m^{-2} h^{-1}kPa^{-1}) values coefficient as estimated for different crops, cited by Carmassi et al. (2013)

Crop	A	B	Growing cond.	Reference
Begonia	0.202	0.026	Angers, France; spr., sum. or aut.; 10-20 pl. m^{-2}; peat-rockwool (pot plants)	Baille et al., (1994a)
Cyclamen	0.318	0.019		
Gardenia	0.533	0.013		
Hibiscus	0.369	0.037		
Impatiens	0.667	0.013		
Pelargonium	0.610	0.017		
Poinsettia	0.120	0.017		
Schefflera	0.608	0.014		
Cucumber	0.24-0.42	0.022-0.038	Almeria, Spain; aut., 2 pl. m^{-2} and spr., 1.33 pl. m^{-2}; perlite	Medrano et al.,(2005)
Gerbera	0.30	0.038	Barcelona, Spain; Jul.-Mar.; 5.7 pl. m^{-2}; Perlite	Marfà et al., (2000)
Rose	0.24	0.017	Volos, Greece; win.; 6 pl. m^{-2}; Perlite	Kittas et al., (1999)
Rose	0.36	0.021	Valencia, Spain; sum.; 7 pl. m^{-2}; Rockwool	Gonzalez-Real (1994)
Tomato	0.58	0.025	Almeria, Spain; aut. and spr.; 7 pl. m^{-2}; perlite bag	Medrano et al. (2004)
Zucchini	0.63	0.007	Viterbo, Italy; aut. and spr.; 2.1 pl. m^{-2}; pumice	Rouphael and Colla (2004)

$$\lambda T = A(1 - \exp(-KLAI))G + BLAIVPD \qquad (5)$$

where 'T' is the measured crop transpiration rate (kg m^{-2} s^{-1}), 'G' the greenhouse solar radiation measured inside (W m^{-2}), 'VPD' the air vapor pressure deficit measured inside (kPa), 'LAI' the calculated leaf area index (m^2 leaf m^{-2} ground), 'K' the light extinction coefficient, 'λ' the vaporization heat of water (J kg^{-1}), and 'A, B' values of equation parameters (A, dimensionless; B, W m^{-2} kPa^{-1}).

However, even in that case, the simplified equation model must firstly be calibrated (i.e., a and b coefficients), under the prevailing environmental conditions. Table 1 shows, corresponding values of a and b model coefficients as calibrated for different crops and cited by Carmassi et al., (2013). It can be observed, that even within the same crop, (i.e., rose) different a and b values were obtained in different environments. However, according to the same author there is no need for model recalibration when higher salinity water was used, based on trials in a semi-closed rockwool cultivation system with gerbera, as transpiration is not affected through stomata closure but through an inhibition of leaf area development.

Estimating crop water needs based on evapotranspiration models according to Merci et al. (2011) indicated lower transpiration values in closed hydroponic systems, compared to open systems as a result of easier buildup of salt in the root zone, which may negatively influence lai. Similarly, Tüzel et al. (2006), reported daily evapotranspiration values of tomato growth in perlite of 4.4 mm in closed and 5.0 mm in open soilless systems.

In any case, the accuracy of predicting the transpiration rate based on evapotranspiration models could be very high, especially when the estimation time interval is based on a few minute basis; therefore it should be recommended especially in high tech greenhouses.

5.3. Lysimeters

Transpiration could be measured directly through the use of devices (e.g., lysimeters) for measuring continual changes in crop mass over short time intervals. The typical structure of a lysimeter, consists of a container to support the growing media with plants, and a weighting device. The structure is in line with the cultivation, but acts as a separate body. The difference between the two measurements, is assumed to be equal to the crop transpiration. According to (Fazlil Ilahi, 2009) a requirement for the perfect measurement with the use of lysimeters, is that the vegetation both inside and immediately outside the lysimeters must be perfectly matched (same height and leaf area index). This requirement has historically not been closely adhered to in a majority of lysimeter studies and has failed to predict real evapotranspiration. Thus, if weighed plants are substantially elevated above surrounding border plants, they are exposed to higher solar radiation and wind, which elevates transpiration values compared to surrounding plants as cited in Beeson (2011).

Lysimeters are considered to be, by multiple authors, the most accurate and quantifiable measurements of transpiration over shorter intervals in real time; and have been used to great success, both for field and greenhouse cultivations and calibrating various evapotranspiration models (Meijer et al., 1985; Beeson, 2011; Schmidt et al. 2013).

6. PLANT BASED ON DEMAND IRRIGATION MONITORING

In order for maintain optimum conditions for the crop growth and production, direct monitoring and evaluating of plant responses to their environment on a real time basis is becoming mandatory especially for high investments cultivations. To date, as we have shown conventional irrigation scheduling approaches, including methods based on measurements of plants climatic environment (e.g., greenhouse temperature, VPD, humidify), monitoring of the substrate water content, or measuring solar radiation; which is related to the plants requirements, through some mathematical or statistical factor. The majority of these methods are time consuming to apply, have restrictions on their application and contain significant sources of experimental error (Gallardo, 1993).

Nowadays, with the great expansion of technology, various plant based sensors have been developed, with the aim to ensuring a more holistic approach on irrigation scheduling, by monitoring physiological changes of the plant itself, instead of its surrounding environment. That allows for us to establish a feedback system in real time based upon controlling the different operational greenhouse functions (e.g., start/stop time of cooling, heating, ventilation) or of evaluating plant responses to irrigation scheduling. According to Ton and Kopyt (2003a), the early detection of plant physiological disorder may be caused by improper irrigation monitoring, allowing adjustment or validation of irrigation regimes such as the volume and timing of irrigation aimed at continuously keeping optimal soil water availability conditions to crops, in relation to a dynamic steady state of the soil-plant-climate system.

Plant sensing technologies have been intensively examined by many researchers, mainly in commercial orchards of grapevines and fruit trees. Except the fact that those crops, generally set fruit, grow and ripen simultaneously, while the irrigation cycle is usually calculated on a weekly basis. In soilless systems, the crop water status is affected in a different way, mainly because of a constant fruit load throughout a large part of the growing season, and the fact that irrigation could be applied multiple times on a daily basis (De Swaef and Steppe, 2010).

Sensors used for plant monitoring, can be broadly classified into two categories. The first category consists of sensors which are in physical contact with the plant (e.g., leaf temperature sensor, stem micro variation sensor) and in some case destructive (e.g., sap flow sensor). In that case, information retrieved on a plants water status obtain from one organ of a single plant may not be a correct representation. In order to the correct respective readings of plant water status, some climate control systems suggest the installation of phyto-sensors in several positions within the greenhouse. The second category consists of sensors which are not plant destructive and usually are at some distance from the crop (e.g., radiometric sensors), while more plants are monitored at the same time.

6.1. Contact Sensors of Plant Monitoring

As cited by Vermeulen et al. (2012), the temperature of the leaf is considered to be one of the most promising and valuable plant responses for automated monitoring of glasshouse crops. Typically, the temperature of a canopy follows a diurnal curve, with day-time temperatures rising, due to an increase in solar radiation and air temperature. In addition, water stressed plants will reduce transpiration and will typically have a higher temperature than a non-stressed plant (DeJonge et al., 2015). Leaf temperature monitoring, has been used as a direct tool to determine the irrigation amount and timing of open field, greenhouses soil cultivations and developing water stress indices (e.g., Crop Water Stress Intex; CWSI). Recently, Naeeni et al. (2014), working with soilless cultivation of four vegetative crops; showed that the index difference between the temperature of the leaf and air showed that leaf temperature which could indicate the amount of water absorbed by the roots and plant transpiration, therefore, it can be used to determined the time of irrigation. Other leaf indices which have been successfully applied for irrigation scheduling in soil cultivations, seemed to be promising for soilless culture, such as the changes of the thickness of a leaf or the measurements of the attenuated pressure of a leaf patch, in response to a constant clamp pressure (Zimmermann et al., 2008; Seeiling et al., 2012).

Indicators of a plants water status, based on the stem diameters micro variation measurements, attracted the highest attention due to their sensitivity and suitability for continuous data recording, which creates the possibility of on-going field monitoring of the crop water status. The stem diameter micro variation, reflects plant water status, and daily increase in stem diameter was found as a reliable criterion for the decision of irrigation time (Lee & Shin, 1998). This enables the possibility of automatic irrigation based directly on the plant itself, rather than on indirect assessments of climatic factors or soil water status (Gallardo et al., 2006). In greenhouse cultivations, the stem diameter varried, depending on the diurnal variation of VPD of atmosphere in the greenhouse and becoming negative each time the VPD rose, indicating that the plants were experiencing water stress conditions. According to the authours' previous work (unpublished), the stem contraction amplitude (i.e., relative water loss) as estimated with the use of stem variation sensor, was higher at 0.448 Kg m^{-2} (Low Irrigation Frequency; LIF) trial, comparing with 0.288 Kg m^{-2} (Medium Irrigation frequency; MIF) trial, in soilless cucumber cultivation, indicating that plants were facing water stress conditions (Figure 1).

According to De Swaef et al. (2010), a modified version of water flow and storage model for trees was applied in soilless tomato stems; indicating that unambiguous determination and interpretation of the plant water status could only be achieved by a simultaneous response of stem variations along with measurements of sap flow. In contrast, the greatest concern when measuring the plant sap flow, is the fact that those methods are rather plant intrusive, which may restrict growth or may cause wounds in the plant and create an entry point for infection. (Ehret et al., 2001).

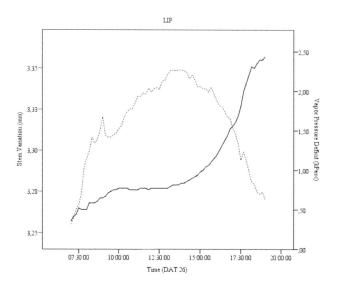

Figure 1. (Continued).

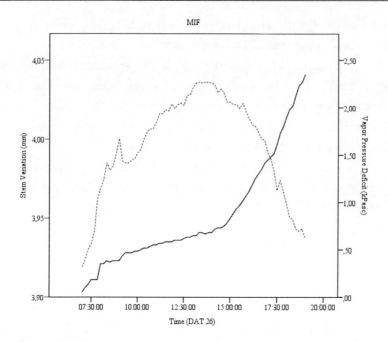

Figure 1. Stem micro variation measurements (mm) (continuous line), and vapor pressure deficit (kPasc) (dotted line), for medium irrigation frequency treatment (MIF; 0.288 Kg m^{-2}) and for low irrigation frequency treatment (LIF; 0.448 Kg m^{-2}).

6.2. Remote Plant Monitoring

Remote sensor readings techniques have been successfully used for years in order to detect water stress and relevant reflectance calculation models have been developed. Despite the fact that few water stress indicators can be estimated remotely, with the use of sensors that do not affect plant development; in the case of greenhouse crops, the remote sensing technology had not been extensively tested. This could be ascribed mainly due to difficulties arising from shadows, resulting from the greenhouse frame and equipment (Kittas et al., 2016).

Remote plant sensors typically measure the reflectance of electromagnetic radiation from crop surfaces. However, the electromagnetic waves which are reflected from the various surfaces differ in length and frequency due to the characteristics of the surface material. It is this difference in reflectance (called the "spectral response") that is measured and can be used to infer crop stress and water requirements (White & Raine, 2008). The most effective spectral indices as cited in (Kittas et al., 2016) are presented in Table 2. The photochemical reflectance index (PRI) and the normalised difference vegetation index (NDVI) are the most commonly used and analyzed indices for crop water stress assessment while the rest indices have being used with either positive or negative results. A brief overview of crop reflectance monitoring is provided by Katsoulas et al. (2016), followed by details on the method of operation, maintenance requirements, typical purchase costs, and the advantages and disadvantages of the method for each method.

Table 2. Reflectance indices for plant water stress detection

Index Reference	Index's calculation	Reference
Photochemical reflectance index; (PRI)	$\dfrac{(R531 - R570)}{(R531 + R570)}$	Sarlikioti et al. (2010); Suarez et al. (2009)
Water intex; (WI)	$\dfrac{(R970)}{(R900)}$	Jones et al. (2004); Clevers et al. (2008)
Normalised difference vegetation index; NDVI(800-680)	$\dfrac{(R800 - R680)}{(R800 + R680)}$	Koksal et al. (2010); Genc et al. (2011)
Normalised difference vegetation index; NDVI(490-620)	$\dfrac{(R490 - R620)}{(R490 + R620)}$	Shimada et al. (2012)
Red normalised difference vegetation index; (rNDVI)	$\dfrac{(R750 - R705)}{(R750 + R705)}$	Kim et al. (2010); Amatya et al. (2012
Modified red edge normalized difference vegetation index; (mrNDVI)	$\dfrac{(R750 - R705)}{(R750 + R705 - 2*445)}$	Kim et al. (2010); Amatya et al. (2012)
Modified red edge simple ratio index; (mrSRI)	$\dfrac{(R750 - R445)}{(R705 + R445)}$	Amatya et al. (2012)
Vogelman red edge index; (VOG REI)	$\dfrac{(R740)}{(R720)}$	Vogelman et al. (1993)

Tsirogiannis et al. 2010, working with portable multispectral radiometer concluded that, the calculated reflectance index of soilless gerbera, with the use of a formula similar to the PRI formula, which indicated an improvement in crop water status after irrigation. In addition, Sarlikioti et al. (2010) who evaluated the use of a photochemical reflectance index (PRI) as an indicator of early water stress, in tomato plants grown in rockwool slabs, concluded that this is possible only when light intensity at crop level is above 700 μmol m^{-2} s^{-1} as at lower values of light intensity, the relation of (PRI) to relative water content was poor in comparison to photosynthesis or fluorescence parameters. Therefore, the use of PRI as a water stress indicator cannot be independent of the ambient light condition. The leaf water index R1300/R1450 which was examined by Seelig et al. (2009), seems to follow the leaf cell relative water content in cowpea and in bean leaves within experiments made in pots under greenhouse conditions and may therefore be used as a feedback-signal for precise irrigation control. Additionally Morales et. al. (2013) indicated that the application of thermal imaging in irrigation could be used as a tool for diagnosis of an early water stress symptoms in soilless cultivation of Philodendron erubescens and Anthurium andraenum under greenhouse conditions in three different tested irrigation volumes (deficit, normal and excess irrigation).

7. DRAIN VOLUMES AND WATER BALANCE METHOD

The amount of nutrient solution outflow from the greenhouse (i.e., drainage), could be applied as a tool for evaluating the irrigation scheduling. In general, by keeping the daily drain volume in the range of 30 to 40% of the nutrient solution it ensures that there will be no build up of salts in the substrate root zone. Lower amounts of drainage solution have also been recorded in experimental findings, as it is affected by the cultivation season, the crop type, the crop length cycle and the type of the substrate.

Tuzel et al. (2004), working with tomatoes cultivated in perlite substrate and a mixture of perlite with peat, indicated that by decreasing the leaching fractions from 15-20% to 5-10%, significant differences in yield between substrates were found only in the spring season compared with the autumn season cultivation. Additionally lower leaching fractions lead to higher salt accumulation at the root environment, resulting in the salinity of drained water increasing in spring.

Measuring the differences in volume, between the water input and the water outflow from a system, could also be used as a method for the indirectly estimating the water uptake (i.e., transpiration rate). This especially applies, if the water uptake is monitored several times on a day-to-day basis, in representative group of plants within the greenhouse, the transpiration could be estimated with high accuracy and could be used as a tool for evaluating the irrigation interval rate and quantity.

CONCLUSION

This chapter has described the most common irrigation approaches used in soilless crop cultivation systems. It has been well documented that conventional irrigation scheduling techniques such as time clock scheduling or irrigation based on solar integrators; when used as a single criterion for irrigation, many times, resulted in inappropriate substrate root conditions, so it is recommended that they are combined. On the other hand, monitoring of the substrate electrical conductivity or volumetric water content, could be used for applying irrigation or as a strategy tool for stimulating vegetative or productive crop growth rate, even though several considerations must be taken into account regarding the reliability of the measurements or the ability to correlate the substrate bulk electrical conductivity to pore water electrical conductivity.

Lastly, the simplified P-M equation transpiration model seems to overcome eliminations of others' transpiration models, therefore it should be recommended for calculating greenhouse crop transpiration rates. However, even in that case there is a need for model calibration, even within the same crop species, as cultivation practices proved to affect the model coefficients. Other methods for directly estimating transpiration, even though they are of high accuracy (i.e., weighting lysimeters) their application is limited mainly for experimental purposes.

In conclusion, there have been tremendous advances in plant monitoring for precision irrigation scheduling in controlled environment agriculture systems. It seems that by monitoring plants, rather than their surrounding environment, it is better understandable the plant's physiological status, allowing for implementing or adjusting of irrigation scheduling. A possible future perspective could therefore include the monitoring of plant water stress indices through the use of remote sensors, utilising non destructive technology, such as the use of reflectance indices on a real time basis.

REFERENCES

Abdel-Razzak, H., Wahb-Allah, M., Ibrahim, A., Alenazi, M. & Alsadon, A. (2016). Response of cherry tomato to irrigation levels and fruit pruning under greenhouse conditions. *J. Agric. Sci. Technol.*, *18*(4), 1091-1103.

Allen, R., Pereira, L., Raes, D. & Smith, M. (1998). Crop Evapotranspiration Guidelines for Computing Crop Water Requirements. FAO Irrigation and Drainage Paper, 56, Rome, Italy.

Baas, R., Nijssen, H. M. C., van den Berg, T. J. M. & Warmenhoven, M. G. (1995). Yield and quality of carnation (Dianthus caryophyllus L.) and gerbera (Gerbera jamesonii L.) in a closed nutrient system as affected by sodium chloride. *Sci. Hortic*, *61*(3-4), 273-284.

Bacci, L., Battista, P. & Rapi, B. (2008). An integrated method for irrigation scheduling of potted plants. *Sci. Hortic.*, *116*(1), 89-97.

Baille, M., Baille, A. & Laury, J. C. (1994). A simplified model for predicting evapotranspiration rate of nine ornamental species vs. climate factors and leaf area. *Sci. Hortic.*, *59*(3-4), 217-232.

Beeson, Jr. R. C. (2011). Weighing lysimeter systems for quantifying water use and studies of controlled water stress for crops grown in low bulk density substrates. *Agric. Water Manag.*, *98*(6), 967-976.

Carmassi, G., Bacci, L., Bronzini, M., Incrocci, L., Maggini, R., Bellocchi, G., Massa, D. & Pardossi, A. (2013). Modelling transpiration of greenhouse gerbera (Gerbera jamesonii H. Bolus) grown in substrate with saline water in a Mediterranean climate. *Sci. Hortic.*, 156, 9-18.

De Graaf, R. & Esmeijer, M. H. (1998). Comparing calculated and measured water comsumption in a study of the (minimal) transpiration of cucumbers grown on rockwool. *Acta Hort.*, *458*, 103-112.

De Swaef, T. & Kathy, S. (2010). Linking stem diameter variations to sap flow, turgor and water potential in tomato Funct. *Plant Biol.*, *37*, 429-438.

DeJonge, K. C., Taghvaeian, S., Trout, T. J. & Comas, L. H. (2015). Comparison of canopy temperature-based water stress indices for maize. *Agric Water Manag.*, *156*, 51-62.

Ehret, D. L., Lau, A., Bittman, S., Lin, W. & Shelford, T. (2001). Automated monitoring of greenhouse crops. *Agronomie.*, *21*(4), 403-414.

Fazlil Ilahi, W. F. (2009). Evapotranspiration models in greenhouse. M.Sc. Thesis, Irrigation and Water Engineering Group, Wageningen University.

Fernández, M. D., Bonachela, S., Orgaz, F., Thompson, R., López, J. C., Granados, M. R., Gallardo, M. & Fereres, E. (2010). Measurement and estimation of plastic greenhouse reference evapotranspiration in a Mediterranean climate. *Irrig. Sci.*, *28*(6), 497-509.

Fields, J. S., Owen, J. S., Zhang, L. & Fonteno, W. C. (2016). Use of the evaporative method for determination of soilless substrate moisture characteristic curves. *Sci. Hortic.*, *211*, 102-109.

Folegatti, M. V., Casarini, E. & Blanco, F. F. (2001). Greenhouse irrigation water depths in relation to rose stem and bud qualities. *Scientia Agricola.*, *58*, 465-468.

Fulcher, A. F., Buxton, J. W. & Geneve, R. L. (2012). Developing a physiological-based, on-demand irrigation system for container production. *Sci. Hortic.*, *138*, 221-226.

Gallardo, M., Thompson, R. B., Valdez, L. C. & Fernandes, M. D. (2006). Use of stem diameter variations to detect plant water stress in tomato. *Irrig. Sci.*, *24*, 241-255.

Gallardo, T. I. (1993). Using Infrared Canopy Temperature and Leaf Water Potential for Irrigation Scheduling in Peppermint (Mentha piperita L.) M.Sc. Thesis, Oregon State University.

Helmer, T., Ehret, D. L. & Bittman, S. (2005). CropAssist, an automated system for direct measurement of greenhouse tomato growth and water use. *Comput. Electron. Agric.*, *48*, 198-215.

Imtiyaz, M., Mgadla, N. P., Chepete, B. & Manase, S. K. (2000). Response of six vegetable crops to irrigation schedules. *Agric Water Manag.*, *45*(3), 331-342.

Incrocci, L., Marzialetti, P., Incrocci, G., Di Vita, A., Balendonck, J., Bibbiani, C., Spagnol, S. & Pardossi, A. (2014). Substrate water status and evapotranspiration irrigation scheduling in heterogenous container nursery crops. *Agric. Water Manag.*, *131*, 30-40.

Jones, H. G. & Tardieu, F. (1998). Modelling water relations of horticultural crops: a review. *Sci. Hortic.*, *74*(1-2), 21-46.

Katsoulas, N. & Kittas, C. (2011). Greenhouse Crop Transpiration Modelling, Evapotranspiration - From Measurements to Agricultural and Environmental Applications, Dr. Giacomo Gerosa (Ed.), ISBN: 978-953-307-512-9, InTech, Available from: http:// www.intechopen.com/books/evapotranspiration-from-measurements-to-agricultural-and-environmental-applications/greenhouse-crop-transpiration-modelling.

Katsoulas, N., Elvanidi, A., Ferentinos, K. P., Kacira, M., Bartzanas, T. & Kittas, C. (2016). Crop reflectance monitoring as a tool for water stress detection in greenhouses: A review. *Biosyst. Eng.*, *151*, 374-398.

Katsoulas. N., Kittas, C., Dimokas, G. & Lykas, C. (2006). Effect of irrigation frequency on rose flower production and quality. *Biosyst. Eng.*, *93*(2), 237-244.

Kittas, C. (1990). Solar radiation of a greenhouse as a tool to its irrigation control. *Int. J. Energ. Res.*, *14*, 881-892.

Kittas, C., Elvanidi, A., Katsoulas, N., Ferentinos, K. P. & Bartzanas, T. (2016). Reflectance indices for the detection of water stress in greenhouse tomato (*Solanum lycopersicum*). *Acta Hort. (ISHS)*, *1112*, 63-70.

Kittas, C., Katsoulas, N. & Baille, A. (1999). Transpiration and canopy resistance of greenhouse soilless roses: measurements and modeling. *Acta Hort.*, *507*, 61-68.

Lee, A. (2009). Understanding Substrate Design. Practical Hydroponics & Greenhouses. November/December, 56-59.

Lee, B. W. & Shin, J. H. (1998). Optimal irrigation management system of greenhouse tomato based on stem diameter and transpiration monitoring. *Agric. Inf. Tech. Asia Oceania.*, 87-90.

Liu, H. J., Cohen, S., Tanny, J., Lemcoff, J. H. & Huang, G. (2008). Estimation of banana (Musa sp.) plant transpiration using a standard 20 cm pan in a greenhouse. *Irrig. Drain Syst.*, *22*(3-4), 311-323.

Lizarraga, A., Boesveld, H., Huibers, F. & Robles, C. (2003). Evaluating irrigation scheduling of hydroponic tomato in Navarra, Spain. *Irrig. Drain.*, *52*(2).

Magán, J. J., Gallardo, M., Thompson, R. B. & Lorenzo, P. (2008). Effects of salinity on fruit yield and quality of tomato grown in soil-less culture in greenhouses in Mediterranean climatic conditions. *Agric. Water Manag.*, *95*(9), 1041-1055.

Mahjoor, F., Ghaemi, A. A. & Golabi, M. H. (2016). Interaction effects of water salinity and hydroponic growth medium on eggplant yield, water-use efficiency, and evapotranspiration. *Inter. Soil and Water Conser. Res.*, *4*, 99-107.

Massa, D., Incrocci, L., Maggini, R., Carmassi, G., Campiotti, C. & Pardossi, A. (2010). Strategies to decrease water drainage and nitrate emission from soilless cultures of greenhouse tomato. *Agric. Water Manag.*, *97*(7), 971-980.

Medrano, E., Lorenzo, P., Sánchez-Guerrero, M. C. & Montero, J. I. (2005). Evaluation and modelling of greenhouse cucumber-crop transpiration under high and low radiation conditions. *Sci. Hortic.*, *105*(2), 163-175.

Meijer, J., Bot, G. P. A., Stanghellini, C. & Cate, A. J. U. (1985). Development and application of a sensitive, high precision weighting lysimeter for use in greenhouses. *J. Agr. Eng. Res.*, *32*, 321-336.

Meric, M. K., Tuzel, I. H., Tuzel, Y. & Oztekin, G. B. (2011). Effects of nutrition systems and irrigation programs on tomato in soilless culture. *Agric. Water Manag.*, *99*(1), 19-25.

Montero, J. I., Antón, A., Muñoz, P. & Lorenzo, P. (2001). Transpiration from geranium grown under high temperatures and low humidities in greenhouses. *Agric. For Meteorol.*, *107*(4), 323-332.

Morales, I., Alvaro, J. E. & Urrestarazu, M. (2013). Contribution of thermal imaging to fertigation in soilless culture. *J. Therm. Anal. Calorim.*, 1-7.

Morille, B., Migeon, C. & Bournet, P. E. (2013). Is the Penman-Monteith model adapted to predict crop transpiration under greenhouse conditions? *Application to a New Guinea Impatiens crop. Sci. Hortic.*, *152*, 80-91.

Murray, J. D., Lea-Cox, J. D. & Ross, D. S. Time domain reflectometry accurately monitors and controls irrigation water applications in soilless substrates. *Acta Hortic.*, *633*, 75-82.

Naeeni, A. E., Esfahani, E. M., Harchegani, M. B., Jafarpour, M. & Golabadi, M. (2014). Leaf Temperature as an Index to Determine the Irrigation Interval. *Plant Physiol.*, 9/1, 89-95

Nebauer, S. G., Sánchez, M., Martínez, L., Lluch, Y., Renau-Morata, B. & Molina, R. V. (2013). Differences in photosynthetic performance and its correlation with growth among tomato cultivars in response to different salts. *Plant Physiol Biochem.*, *63*, 61-69.

Nemali, K. S., Montesano, F., Dove, S. K. & Marc, W. van Iersel. (2007). Calibration and performance of moisture sensors in soilless substrates: ECH2O and Theta probes. *Sci. Hortic.*, *112*(2), 227-234.

Oki, L. R. & Lieth, J. H. (2008). Irrigation in soilless production, In: Raviv, M. & Lieth, J.H (Ed.): Soilless Culture: Theory and practice (first edition). Elsevier, London, pp. 117-155.

Orgaz, F., Fernández, M. D., Bonachela, S., Gallardo, M. & Fereres, E. (2005). Evapotranspiration of horticultural crops in an unheated plastic greenhouse. *Agric. Water Manag.*, *72*(2), 81-96.

Pardossi, A., Incrocci, L., Incrocci, G., Fernando, B., Bacci, P., Rapi, L., Marzialetti, B., Hemming, P. & Balendonck J. (2009). Root Zone Sensors for Irrigation Management in Intensive Agriculture. *Sensors.*, *9*(4), 2809-2835.

Pollet, S., Bleyaert, P. & Lemeur, R. (2000). Application of the Penman-Monteith model, to calculate the evapotranspiration of head lettuce (Lactuca Sativa L. Var. Capitata) in glasshouse conditions. *Acta Hort.*, *519*, 151-162.

Raviv, M., Wallach, R., Silber, A., Medina, S. & Krasnovsky, A. (1999). The effect of hydraulic characteristics of volcanic materials on yield of roses grown in soilless culture. *J. Am. Soc. Hortic. Sci.*, *124*(2), 205-209.

Rouphael, Y. & Colla, G. (2005a). Growth, yield, fruit quality and nutrient uptake of hydroponically cultivated zucchini squash as affected by irrigations systems and growing seasons. *Sci. Hort.*, *105*, 177–195.

Rouphael, Y., Cardarelli, M., Rea, E. & Colla, G. (2008). The influence of irrigation system and nutrient solution concentration on potted geranium production under various conditions of radiation and temperature. *Sci. Hortic.*, *118*(4), 328-337

Saha, U. K., Papadopoulos, A. P., Hao, X. & Khosla, S. (2008). Irrigation strategies for greenhouse tomato production on rockwool. *HortScience.*, *43*(2), 484-493.

Sarlikioti, V., Meinen, E. & Marcelis, L. F. M. (2011). Crop Reflectance as a tool for the online monitoring of LAI and PAR interception in two different greenhouse Crops. *Biosyst Eng.*, *108*(2), 114-120.

Savvas, D. & Lenz, F. (2000). Effects of NaCl or nutrient-induced salinity on growth, yield, and composition of eggplants grown in rockwool. *Sci. Hortic.*, *84*(1-2), 37-47.

Savvas, D., Stamati, E., Tsirogiannis, I. L., Mantzos, N., Barouchas, P. E., Katsoulas, N. & Kittas, C. (2007). Interactions between salinity and irrigation frequency in greenhouse pepper grown in closed-cycle hydroponic systems. *Agr. Water Manag.*, *91*, 102-111.

Schmidt, C. D. S., Pereira, F. A. C., Oliveira, A. S., Gomez Júnior, J. F. G. & Vellame, L. M. (2013). Design, installation and calibration of a weighing lysimeter for crop evapotranspiration studies. *Water Res. and Irrig. Manag.*, *2*, 77-85.

Schröder, F. G. & Lieth, J. H. (2002). Irrigation control in hydroponics. In Hydroponic Production of Vegetables and Ornamentals, Savvas, D., Passam, H., ed. Embryo Publishing, Athens, Greece, pp. 265-296.

Seelig, H. D., Hoehn, A., Stodieck, L. S., Klaus, D. M., Adams, W. W. & Emery, W. J. (2009). Plant water parameters and the remote sensing R 1300/R 1450 leaf water index: Controlled condition dynamics during the development of water deficit stress. *Irrig Sci.*, *27*(5), 357-365.

Seelig, H. D., Stoner, R. J. & Linden, J. C. (2012). Irrigation control of cowpea plants using the measurement of leaf thickness under greenhouse conditions. *Irrig. Sci.*, *30*(4), 247-257.

Sezen, S. M., Celikel, G., Yazar A., Tekin, S. & Kapur, B. (2010). Effect of irrigation management on yield and quality of tomatoes grown in different soilless media in a glasshouse. *Scientific Research and Essays*, *5*(1), 041-048.

Shin, J. H., Park, J. S. & Son, J. E. (2014). Estimating the actual transpiration rate with compensated levels of accumulated radiation for the efficient irrigation of soilless cultures of paprika plants. *Agric. Water Manag.*, *135*, 9-18.

Silber, A., Xu, G. & Wallach, R. (2003). High irrigation frequency: The effect on plant growth and on uptake of water and nutrients. *Acta Hortic.*, 2003, *627*(3), 89-96.

Şimşek, M., Tonkaz, T., Kaçira, M., Çömlekçioğlu, N. & Doğan, Z. (2005). The effects of different irrigation regimes on cucumber (Cucumbis sativus L.) yield and yield characteristics under open field conditions. *Agric. Water Manag.*, *73*(3), 173-191.

Steppe, K., De Pauw, D. J. W. & Lemeur, R. (2008). A step towards new irrigation scheduling strategies using plant-based measurements and mathematical modelling. *Irrig Sci.*, *26*(6), 505-517.

Ton, Y. & Kopyt, M. (2003a). Phytomonitoring: A Bridge from Sensors to Information Technology for Greenhouse Control. *Acta Horticulturae.*, *614*, 639-644.

Tsirogiannis, I., Katsoulas, N. & Kittas, C. (2010). Effect of irrigation scheduling on gerbera flower yield and quality. *HortSci.*, *45*(2), 265-270.

Tüzel, I. H., Meric, K. M. & Tüzel, Y. (2006). Crop Coefficients in Simplified Hydroponic Systems. *ISHS Acta Hort.*, *719*, 551-556.

Tüzel, I. H., Tüzel, Y., Gül, A. & Eltez, R. Z. (2004). Effects of different leaching fractions and substrates on tomato growing. *Acta Hort. (ISHS)*, *633*, 301-308.

Tüzel, I. H., Tüzel, Y., Gül, A. & Eltez, R. Z. (2004). Effects of different leaching fractions and substrates on tomato growing. *Acta Hort. (ISHS)*, *633*, 301-308.

Valdés, R., Miralles, J., Franco, J. A., Sánchez-Blanco, M. J. & Bañón, S. (2014). Using soil bulk electrical conductivity to manage saline irrigation in the production of potted poinsettia. *Sci. Hortic.*, *170*, 1-7.

Vermeulen, K., Aerts, J. M., Dekock, J., Bleyaert, P. & Berckmans, D., Steppe, K. (2012). Automated leaf temperature monitoring of glasshouse tomato plants by using a leaf energy balance model. *Comput Electron Agric.*, *87*, 19-31.

Villarreal-Guerrero, F., Kacira, M. & Fitz-Rodríguez, E. (2012). Simulated performance of a greenhouse cooling control strategy with natural ventilation and fog cooling. *Biosyst. Eng.*, *111*(2), 217-228.

Wallach, R. (2008). Physical characteristics of soilless media, In: Raviv, M. & Lieth, J.H (Ed.): Soilless Culture: Theory and practice (first edition). Elsevier, London, pp. 41-108.

White, S. & Raine, S. R. (2008). A grower guide to plant based sensing for irrigation scheduling. National Centre for Engineering in Agriculture Publication. 1001574/6, USQ, Toowoomba.

Yang, X., Short, T. M., Fox, R. D. & Bauerle, W. L. (1990). Transpiration, leaf temperature and stomatal resistance of a greenhouse cucumber crop. *Agric. Forest Meteorol.*, *51*, 197-209.

Zhao, S., Wang, Q., Yao Y, Du, S., Zhang, C., Li, J. & Zhao, J. (2016). Estimating and validating wheat leaf water content with three MODIS spectral indexes: A case study in Ningxia Plain, *China. J. Agric. Sci. Technol.*, *18*(2), 387-398.

Zimmermann, D., Reuss, R., Westhoff, M., Gessner, P., Bauer, W., Bamberg, E., Bentrup, F. W. & Zimmermann, U. (2008). A novel, non-invasive, online-monitoring, versatile and easy plant-based probe for measuring leaf water status. *J Exp. Bot.*, *59*(11), 3157-3167.

In: Advances in Hydroponics Research
Editor: Devin J. Webster
ISBN: 978-1-53612-131-5
© 2017 Nova Science Publishers, Inc.

Chapter 2

NITRATE-AMMONIUM RATIOS AND SILICON USED IN HYDROPONICS

Renato de Mello Prado, Rafael Ferreira Barreto and Guilherme Felisberto
São Paulo State University (UNESP),
School of Agricultural and Veterinary Sciences, Jaboticabal, Brazil

ABSTRACT

Excess nitrate in the nutrient solution of some hydroponic crop species has attracted the attention of regulatory agencies because this form of nitrogen can be often toxic to humans, especially children. Inclusion of ammonium in hydroponic nutrient solutions can promote significant increase in productivity and thus, help meet the growing demand for safer foods. However, several factors may alter the availability of ammonium to plants, which may lead to phytotoxicity. However, silicon can mitigate abiotic stresses such as ammonium phytotoxicity, and can be used to prevent possible damage to plants from this form of nitrogen. Therefore, the objective of this review was to gather research information about the nitrogen forms used in hydroponics, their relationships and peculiarities, as well as information on the use of silicon, its available sources and role in mitigating the toxic effects of ammonium.

1. Introduction

Plant-available nitrogen (N) in hydroponic crops can be found in the form of nitrate (NO_3^-) and ammonium (NH_4^+) ions. To be incorporated into organic compounds in plants, N must be in the form of NH_4^+. When NO_3^- is absorbed by the plant, assimilatory reduction, one of the largest energy consumption process in plants, must occur (Salsac et al., 1987). Thus, the lower metabolic energy expenditure required by the plant to utilize ammonium nutrition can be converted to higher dry matter production. However, an optimal ratio of NO_3^-/NH_4^+ provides greater growth in most species, than does N provided in only one form (Britto & Kronzucker, 2002).

Furthermore, growing plants in nutrient solution which have their N content in the form of NO_3^- can lead to symptoms of deficiency of some micronutrients such as iron (Fe), due to the high pH value on the surface of roots (Nikolic & Römehld. 2003). Ammoniacal nutrition, in contrast, increases the availability of Fe by acidifying the rhizosphere, increasing Fe absorption rate, and alleviating the deficiency symptoms of this micronutrient (Prado & Esteban, 2011).

As a result of climate change, NH_4^+ may become more important for plant nutrition. In conditions of increased atmospheric CO_2, there is a reduction in photorespiration, nitrite intake in chloroplasts, and competition for ferredoxin reduced from nitrate reductase activity in the chloroplast stroma, and thus, a decrease in nitrate assimilation in leaves, as demonstrated by Bloom et al. (2014) in wheat.

In *Arabidopsis thaliana*, the concentration of NH_4^+ in cells is regulated by phosphorylation of carrier proteins, and active absorption stops when the optimum concentration is reached. However, when the concentration of NH_4^+ in cells rapidly increases, which can occur under certain environmental conditions, carrier phosphorylation cannot occur rapidly enough to avoid excessive accumulation of NH_4^+, leading to phytotoxicity (Lanquar & Frommer, 2010). Moreover, NH_4^+ can be absorbed by the aquaporins, non-selective cation transporters, and potassium conveyors, which increases the risk of toxicity (Bittsánszky et al., 2015).

A common feature of plants grown using NH_4^+ is the increased production of reactive oxygen species (ROS). With moderate NH_4^+ supply, this can sometimes be beneficial by activating the plant defense system, making it more resistant to attack by pathogens. With excess NH_4^+, however, it can be harmful causing foliar chlorosis by the degradation of chloroplasts (Liu & Wirén, 2017).

In this context, the GENPLANT-Group of Studies in Plant Nutrition of UNESP (São Paulo State University), Brazil, has included research on the use of silicon (Si), a beneficial element known to reduce abiotic stress, in its work on ammonium nutrition (Epstein, 2001). With respect to toxicity of NH_4^+, as shown by the increase in ROS, Si acts by increasing the activity of anti-oxidant enzymes, that are part of the plant defense mechanism, increasing the ability of the plant to remove ROS. Thus, damage to the cell membrane is decreased, making the plant less sensitive to NH_4^+ toxicity (Gao et al., 2014).

Note that the ability of Si to alleviate NH_4^+ toxicity depends on several factors, including the concentration of NH_4^+ in the nutrient solution. The effects of Si are more evident in alleviating toxicity, such that the decrease in dry matter production is no more than 10% (Barreto et al., 2016). In addition, it is important that Si is provided before NH_4^+, so that the plant can absorb it before experiencing abiotic stress, and therefore, tolerate the NH_4^+ induced stress.

The beneficial effects of Si on relieving the toxicity of NH_4^+ becomes more evident when using ratios of NO_3^-/NH_4^+ (Silva Júnior et al., 2015) than when using the isolated form. In cucumber grown using NO_3^-/NH_4^+, although Si alleviated the NH_4^+ toxicity, dry matter of both crops was lower than respective plants growing on NO_3^- medium (Campos et al., 2016).

In this context, results indicate that despite the ability of Si to alleviate the NH_4^+ toxicity, further research is necessary to ensure higher increase in productivity. However, at acceptable concentrations of NH_4^+ in the nutrient solution, the addition of Si can prevent possible cases of phytotoxicity, depending on factors that may increase the uptake of ammonium by plants.

2. Nitrate-Ammonium Ratios and Ammoniacal Phytotoxicity in Hydroponics

For hydroponic crops, the addition of N occurs only through nitrogen fertilizers manufactured by industries that require large amounts of energy to produce the high temperatures and pressures required to break the strong triple bond of atmospheric N_2 and form ammonia (NH_3) (Table 1).

Among nitric and ammoniacal fertilizers, calcium nitrate and potassium nitrate are most often used to supply N in hydroponics. These possess considerable amounts of Ca and K, which are also required in large amounts by crops, as well as N in the form of NO_3^- (Table 1).

Another factor that encourages the use of nitrate sources is the increased productivity of crops, and an absence of toxicity risk, even with high concentrations of this form of N in leaf tissue, as reported in lettuce, by Fu et al. (2017). However, fertilization with high concentrations of NO_3^- can harm some quality characteristics of the crop, such as synthesis of vitamins (Albornoz, 2016). In addition, a major concern regarding the consumption of foods with high NO_3^- levels is the increased risk of methemoglobinemia, especially in children. According to the Scientific Committee on Food (SCF), the daily intake limit of NO_3^- is 3.65 mg kg^{-1} body weight. Thus, the European Union encourages production of foods with lower NO_3^- content, and limits NO_3^- levels in some foods according to the type of cultivation and growing season (Table 2).

An alternative to the reduction of NO_3^- content in hydroponic crops is to provide N in the form of NH_4^+. However, ammoniacal nitrogen sources, when used for the preparation of nutrient solution on a commercial scale, do not usually constitute more than 20% of the total N concentration. Most of the N is provided as NO_3^- owing to the risk of NH_4^+ toxicity on accumulation of this cation in plant cells (Bittsánszky et al., 2015).

Some species, such as snap bean, are very sensitive to ammonium nutrition in the initial stage of development, and exhibit characteristic symptoms of toxicity, even at low concentrations of NH_4^+ in the nutrient solution (Barreto et al., unpublished data) (Figure 1).

Table 1. Fertilizers containing nitrogen that are commonly used in the preparation of nutrient solutions

Fertilizer	Formula	Nutrient	Concentration (%)	Solubility (g L^{-1} at 20°C)
Potassium nitrate	KNO_3	K	38	316
		$N-NO_3^-$	13	
Calcium nitrate	$Ca(NO_3)_2$ $4H_2O$	Ca	19	1299
		$N-NO_3^-$	15,5	
Magnesium nitrate	$Mg(NO_3)_2$ $6H_2O$	Mg	9	760
		$N-NO_3^-$	11	
Ammonium nitrate	NH_4NO_3	$N-NO_3^-$	16,5	1920
		$N-NH_4^+$	16,5	
Ammonium sulphate	$(NH_4)_2SO_4$	S	24	754
		$N-NH_4^+$	21	
Monoammonium phosphate	$NH_4H_2PO_4$	P	26	365
		$N-NH_4^+$	12	
Ammonium chloride	NH_4Cl	Cl	66	546
		$N-NH_4^+$	26	

Table 2. Maximum levels of NO_3^- allowed in fresh vegetables by European Union Regulation (EC) No. 1881/2006, amended in 2011

Product	Specifics	Maximum level NO_3^- (mg kg^{-1} F.W.)
Fresh spinach (*Spinacea oleracea*)		3500
Preserved, deep-frozen or frozen spinach		2000
Fresh lettuce (*Lactuca sativa*)	Winter crop:	
	Grown under cover	5000
	Grown in the open air	4000
	Summer crop:	
	Grown under cover	4000
	Grown in the open air	3000
"Iceberg" type lettuce	Grown under cover	2500
	Grown in the open air	2000
Rucola (*Eruca sativa, Diplotaxis* sp., *Brassica tenuifolia, Sisymbrium tenuifolium*)	Winter crop	7000
	Summer crop	6000

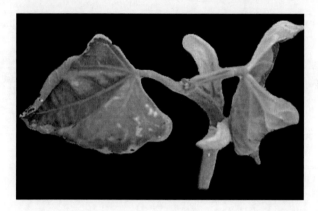

Figure 1. Snap bean ammonium toxicities symptoms in early stage of development induced by 0.75 mmol L^{-1} NH_4^+ in the nutrient solution applied to substrate. Photo: Barreto, 2016 (not published).

In previous research on NH_4^+ toxicity, the most commonly used N sources were ammonium sulfate (Vega-Mas et al., 2015; Balkos et al., 2010; Ariz et al., 2011) and ammonium chloride (Campos et al., 2015; Barreto et al., 2016), sources that have considerable amounts of accompanying S and Cl ions (Table 1). Therefore, to investigate whether these accompanying ions affected NH_4^+ toxicity, we set up an experiment to compare these two N sources (unpublished data); the results showed that the S and Cl ions had no effect on NH_4^+ toxicity induced by 15 mmol L^{-1} NH_4^+, in 'Micro-Tom' tomato plants.

In this context, knowledge of the most beneficial NO_3^-/NH_4^+ ratios, as well as the factors that may affect plant growth when using these ratios, is fundamental to ensure that ammonium nutrition is provided in a way which is beneficial to plant growth.

3. FACTORS AFFECTING NITRATE-AMMONIUM RATIOS

3.1. Air and Nutrient Solution Temperature

At high temperatures, the higher rate of transpiration facilitates greater contact of N with the roots and, consequently, its absorption. In such conditions, even with a low ratio of NH_4^+ to NO_3^-, large amounts of NH_4^+ can be absorbed. This may explain the findings of Liu et al. (2017), in which tomato plants grown at a temperature between 14–30°C tolerated up to 2 mmol L^{-1} NH_4^+, while plants grown from 5–18°C showed tolerance to 4 mmol L^{-1} NH_4^+. In a study investigating ammonium nutrition, severe symptoms of NH_4^+ toxicity were observed in cauliflower within hours of a rapid increase in greenhouse air temperature from 30°C to approximately 40°C (Barreto et al., unpublished data) (Figure 2).

3.2. Light Irradiation

There is a strong relationship between light irradiation levels and nitrogen nutrition, because the photosynthetic rate and the availability of C may be limited in plants cultivated under low irradiation levels. Ariz et al. (2011) reported that pea plants grown under low light irradiation levles (300 µmol photon m^{-2} s^{-1}) were unable to regulate their internal NH_4^+ content, leading to toxicity. Plants grown under high light irradiation (700 µmol photon m^{-2} s^{-1}) and high C availability, could tolerate high concentrations of NH_4^+ in the external medium. Authors point out that the additional C obtained at higher light irradiation levels was not used to increase NH_4^+ uptake, but rather as a source of energy used to maintain low NH_4^+ content in their tissues. Thus, one practical implication of these results for hydroponic cultivation is that extended periods of cloudy weather may cause crops to become the more sensitive NH_4^+.

3.3. pH of the Nutrient Solution

While the pH value of the nutrient solution remains in the range from 2 to 7, the NH_4^+ ion remains in its stable form. At pH values above 7, the concentration of NH_4^+ decreases, and that of ammonia increases (De Rijck & Schrevens, 1999). However, considering that the pH value of hydroponic solution is maintained between 5 and 6 to ensure the high availability of all nutrients, high ammonia concentrations are not expected to occur in hydroponic systems.

3.4. Potassium Concentration in Nutrient Solution

Although the transporter protein involved in N absorption, which becomes phosphorylated and closes when it reaches the optimum content of NH_4^+ cells (Lanquar & Frommer, 2010), NH_4^+ can be absorbed by aquaporins, non-selective cation transporters and potassium transporters (Bittsánszky et al., 2015). Accordingly, Balkos et al. (2010) found that very low concentrations of K in the nutrient solution (0.1 mmol L^{-1}) favor the absorption of NH_4^+, whereas lower NH_4^+ absorption occurs with increased K concentration (5 mmol L^{-1}).

Figure 2. Cauliflower ammonium toxicities symptoms induced by 15 mmol L^{-1} NH_4^+ in the nutrient solution applied to the substrate. Severe symptoms appeared few hours after air temperature had risen from approximately 30°C to about 40°C in the greenhouse. Photo: Barreto, 2016 (not published).

4. NITRATE-AMMONIUM RATIOS FOR SOME PLANTS IN HYDROPONICS

Although there was an indication of the best NO_3^-/NH_4^+ ratio for some crops (Table 3), it remains unclear whether these proportions are suitable for other crops of the same species, owing to factors affecting NH_4^+ absorption, as discussed previously. However, information from the literature indicates that NH_4^+ can be harmful to the crop yield when it is the only N form available (Table 3). Moreover, it is evident that having the correct ratio between the two forms of N increases the crop yield. In this sense, adopting measures that can prevent the toxicity of NH_4^+ is a good alternative to its inclusion in hydroponics. In this context, Si is a beneficial element for preventing abiotic stress (Epstein, 2001).

5. SILICON IN HYDROPONICS

Hydroponics is a tool of fundamental importance in the development of plant nutrition. This system makes it possible to conduct nutrient element omission studies, and to study their chemical forms, concentrations, proportions, and interactions between them, facilitating the understanding of their effects in isolation and in combination.

In studies on plants grown in soil, it is not possible to isolate certain factors. They are, however, closer to reality in that soil is a dynamic system in which transformation processes, losses, and gains happen all the time. One of factors that cannot be isolated in soil-based experiments is the presence of Si, because it is found in silicate minerals in the form of silica, which constitutes about 50 to 70% of the soil mass (Ma & Yamaji, 2006).

However, most of the basic studies of plant nutrition have been developed in hydroponic experiments, and the presence of Si has been neglected. This is of concern, because plant responses to nutrients in the absence of Si may be different from their response in its presence in the medium. Therefore, it is important to include Si in the nutrient solution to evaluate plant responses to any factor. This ensures the study can generate nutritional information that is closer to the reality of cultivation, and will increase the benefits of growing and producing crops.

Si is commonly found in plant tissues. It is absorbed in the form of orthosilicic acid (OSA, H_4SiO_4) actively or passively, depending on its concentration and on the plant species (Liang et al., 2006; Apaolaza-Hernandez, 2014). In the soil, concentrations of OSA vary from 0.1 to 0.6 mM (Epstein, 2001), whereas in river water, normally used in hydroponics, a concentration of around of 150 µM would be considered typical (Tréguer et al., 1995).

The use of Si in commercial hydroponic crops is still being developed. However, owing to the potentially small Si concentrations in the water used, its supplementation in the nutrient solution is essential, because the plants develop on inert substrates (Savvas et al., 2007).

Table 3. Plant species, N concentration and the NO_3^-/NH_4^+ ratio corresponding to higher harvested, according to the authors

Plants	[N] mmol L^{-1}	NO_3^-/NH_4^+ ratios	Authors
Swiss chard (*Beta vulgaris*)	7,5	75/25	Barickman & Kopsell (2016)
Tomato (*Solanum lycopersicum*)	23	70/30	Borgognone et al. (2013)
Cole (*Brassica rapa*)	11	100/0; 50/50	Fallovo et al. (2009)
Strawberry (*Fragaria x ananassa*)	14,3	75/25; 50/50	Tabatabaei et al. (2008)
Spinach (*Spinacia oleracea*)	12	75/25	Xing et al. (2015)
Hot pepper (*Capsicum annuum*)	15	100/0; 73/27	Hojin et al. (2013)
Lettuce (*Lactuca sativa*)	6	75/25	Wang & Shen (2011)
Broccoli (*Brassica oleracea*)	3,5	50/50	Zaghdoud et al. (2016)

Plants considered non-accumulators of Si, such as tomato and cucumber, passively absorb silicon, whereas those considered Si accumulators, such as rice, possess specific carrier proteins capable of actively absorbing Si. These carriers are responsible for the xylem reaching Si contents higher than that in the nutrient solution, in minutes (Mitani & Ma, 2005).

In non-accumulating plants, this implies the need either for foliar applications of Si, aiming to transpose the low absorption, or for higher concentrations of Si in the nutrient solution. It also suggests the possibility of using lower Si concentrations when cultivating accumulating plants.

Once absorbed, Si is translocated through xylem to tissues, and is deposited as amorphous silica ($SiO_2.nH_2O$) in the cell wall and intercellular spaces in the outer subcuticular layer of cells. Once there, it becomes immobile and cannot be redistributed in plant (Ma et al., 2011).

Considered a beneficial element, mainly for acting as a mitigator of biotic and abiotic stresses (Ma & Yamaji, 2006), Si has been studied in hydroponic crops as an inducer of tolerance to water deficit and salt stress in tomato (Cao et al., 2015; Haghighi & Pessarakli, 2013), to metal phytotoxicity in barley, cotton, rice, and maize (Liang et al., 2001; Farooq et al., 2013; Ali et al., 2013; Li et al., 2012; Kaya et al., 2009), and to NH_4^+ in tomato, corn, cucumber, and canola (Barreto et al., 2016; Campos et al., 2015; Campos et al., 2016; Gao et al., 2014; Bybordi, 2010).

Crops such as lettuce, cucumber, melon, and tomato, as well as roses and other ornamentals demonstrated superior performance on receiving Si in the nutrient solution, even when they were not subjected to any form of stress (De Kreij et al., 1999; Bae et al., 2010; Lim et al., 2012). This demonstrates that the benefits of Si transcend stress mitigation.

5.1. Silicon Sources for Hydroponics

For hydroponic systems, a limited number of Si sources are available on the market, with potassium silicate being the most commonly used. This source also contains potassium ions, a plant nutrient that plays a key role in the enzymatic activation of the plant and in osmotic regulation (Prado, 2008). It is noted that concentrations of up to 2 mmol L^{-1} of Si (56 mg L^{-1}) in the form of potassium silicate are commonly used in the nutrient solution (pH 5.8 ± 0.1), because Si polymerization occurs at higher concentrations, causing Si to become unavailable to plants, especially at pH values below 9 (McKeague & Cline, 1963).

The order of Si addition in preparation of the nutrient solution may also decrease its availability. In our studies, we obtained higher success levels by adding the Si source in the water first, adjusting pH value to 5.8 (± 0.1), and only then adding the other elements.

A preparation made by mixing two pH adjusted solutions that were two times more concentrated, one containing the required nutrients, and the other containing only Si did not present any problems with Si polymerization. Another factor that decreases the availability of Si is its precipitation with bivalent cations forming silicates of low solubility such as calcium, magnesium, and zinc silicates, among others (Liang et al., 2007). Thus, nutrient solutions with high contents of these elements can favor Si precipitation, reducing Si availability to the plants.

In recent years, other sources of Si have been developed. For the most part, these aim to provide greater amounts of OSA than is found in potassium silicate. After obtaining the OSA, it is necessary to stablilize the OSA molecules with some additives, such as sorbitol, polyethylene glycol, choline, 1,2 propylene glycol, and micronutrients, because OSA tends to agglomerate to form dimers, trimers, and polymers. OSA is obtained by allowing an acid to attack soluble silicates such as potassium and sodium silicate. This can give a stabilized pH value for OSA, ranging from 0.4 to 4, depending on the additive used.

Stabilized OSA sources have low Si concentrations (9–14 g L^{-1}) than potassium silicate has (> 100 g L^{-1}). However, Si of OSA is readily available to plants, requiring applications of a relatively low concentration (0.3 mg L^{-1} Si) to obtain satisfactory improvements in photosynthetic activity and productivity, as reported by Kleiber et al. (2015) in tomato which had been treated with an excess of Mn. In a study by Genplant in cotton, higher production with OSA was observed at 1.66 g L^{-1} Si, whereas the higher production with potassium silicate was at 1.34 g L^{-1} Si, both as a total supply of four applications (Souza Júnior et al., unpublished data).

Genplant studies with stabilized OSA have recommended against the use of concentrations above 0.8 mmol L^{-1} of Si in the nutrient solution (pH 5.8 ± 0.1), considering that Si showed precipitation at this concentration after 72 hours. A solution of stabilized OSA in deionized water (pH 5.8 ± 0.3), used for fertigation or foliar application, OSA was stable at a concentration of 21 mmol L^{-1} (0.6 g L^{-1}) of Si for up to 72 hours. However, there is a need for further research to understand and possibly increase the stability of these sources, for their use in nutrient solutions, and solutions with other compounds, to diversify and optimize Si application.

5.2. Silicon Mitigates Ammonium Phytotoxicity

High concentrations of NH_4^+ in the nutrient solution promotes deleterious effects in plants. It could induce toxicity by increasing the intracellular pH value, by increasing osmotic and nutritional imbalances, and by causing hormonal disorders (Gerendás et al., 1997). It can also lead to a large decrease in chlorophyll concentration, compromising the photosynthetic rate (Wang et al., 2010; Su et al., 2012). This can cause accumulation of ROS, leading to oxidative stress (Wang et al., 2010), and will also reduce stomatal conductance and transpiration (Lopes & Araus, 2006). Some of these negative effects are not related to the nutritional role of ammonium, and therefore, has recently been attributed to its function in cellular signaling (Liu & Wirén, 2017).

Ammonium ion induces toxicity in plants as it reaches concentrations at which plants cannot assimilate it. An alternative to minimize its phytotoxic effects would be to increase the ability of plants to assimilate this form of nitrogen. In this regard, Campos et al. (2015) reported that in hydroponic maize, supplying silicon (10 mM) to the plants increased the N content of the plants, despite the toxic concentration of NH^+_4 used (Table 4).

Table 4. Accumulation of nitrogen and silicon in the shoots of maize plants submitted to ammonium comcentrations in the absence and presence of silicon (Campos et al., 2015)

NH_4^+ concentration	N accumulation in the shoots		Si accumulation in the shoots	
	Without Si	With Si	Without Si	With Si
	------------ mg per plant ------------			
30 mM	52.49 bB	104.08 aA	7.86 bB	19.22 aA
60 mM	60.91 aB	86.38 bA	9.31 aB	16.15 bA
	------------ F-values ------------			
N	6.25*		4.46^{ns}	
Si	429.26**		439.30**	
NxSi	49.32**		27.17**	
MSD[1]	5.49		1.28	
CV%	6.0		8.1	

Identical upper case letters on the lines do not differ by the Tukey test ($P < 0.05$). Identical lower case letters in the columns do not differ by the Tukey test ($P < 0.05$). *, ** and ^{ns}: significant ($P < 0.05$), ($P < 0.01$) and non-significant, respectively. [1] Minimum significant difference.

Table 5. Dry matter of the shoots and roots of maize plants submitted to ammonium concentrations in the absence and presence of silicon (Campos et al., 2015)

NH_4^+ concentration	Shoot dry matter		Root dry matter	
	Without Si	With Si	Without Si	With Si
	------------ g per plant ------------			
30 mM	1.667 aB	2.878 aA	0.298 aB	1.057 aA
60 mM	1.409 bB	2.011 bA	0.265 bB	0.414 bA
	------------ F-values ------------			
N	169.06**		753.79**	
Si	440.17**		1357.54**	
NxSi	49.70**		612.54**	
MSD[1]	0.13		0.04	
CV%	5.3		5.9	

Identical upper case letters on the lines do not differ by the Tukey test ($P < 0.05$). Identical lower case letters in the columns do not differ by the Tukey test ($P < 0.05$). * and ** ($P < 0.05$) and ($P < 0.01$), respectively. [1] Minimum significant difference.

Table 6. Effects of the interactions between varieties, silicon and nitrogen for height, leaf area and accumulation of dry matter mass (DM) for two varieties of *Cucumis sativus* plants (Campos et al., 2016)

Varieties	Height (cm)		Leaf area (cm^2)		DM (g per plant)	
	NH_4^+	NO_3^-	NH_4^+	NO_3^-	NH_4^+	NO_3^-
Hokushin	66.42 bB	109.83 aA	578.90 bB	1400.26 aA	1.95 bB	4.61 aA
Tsubasa	83.08 aB	99.00 bA	710.28 aB	1298.14 bA	2.55 aB	4.16 bA
Silicon						
0 mmol L^{-1}	49.13 bB	103.75 aA	394.07 bB	1342.69 aA	1.36 bB	4.34 aA
1 mmol L^{-1}	87.25 aB	104.75 aA	768.20 aB	1345.35 aA	2.70 aB	4.37 aA
10 mmol L^{-1}	87.88 aB	104 4 aA	771.51 aB	1359.55 aA	2.69 aB	4.44 aA

Equal lower-case letters in the columns do not differ by the Tukey test (P < 0.05). Equal upper-case letters on the lines do not differ by the Tukey test (P < 0.05).

These authors attributed the greater accumulation of N to the ability of Si to increase the activity of enzymes responsible for the reduction and assimilation of N, and to metabolic processes linked to its assimilation during photosynthesis.

At the minimum toxic concentration of NH_4^+ (30 mM), in presence of Si, N accumulation almost doubled, whereas at higher concentrations, the increase did not reach 50% (Table 4). In addition, Campos et al. (2015) reported that at very high concentrations of NH_4^+ (60 mM) roots are damaged, which affects dry matter accumulation (Table 5) and therefore N accumulation.

These results indicate that the alleviation of NH_4^+ toxicity depends not only on the supply of Si but also on the toxic concentration of NH_4^+. This is because very high N concentrations greatly affect the plant metabolism and development, to the point that Si is not able to play its role as a mitigator of stress.

In a study using lower concentrations of N (10 mmol L^{-1}) in the form of NO_3^- and NH_4^+, three Si concentrations (0, 1 and 10 mmol L^{-1}), and two varieties of cucumber (Hokushin and Tsubasa) grown under hydroponics, Campos et al. (2016) observed that the nitrate did not have a phytotoxic effect on cucumber, and that the mitigating effects of Si on the NH_4^+ phytotoxicity was independent of plant variety. However, in the absence of Si in the nutrient solution, the ammonium supply reduced plant height and leaf area by 30%, and plant dry mass production by approximately 50% (Table 6). It was also observed that the detrimental effects of the high concentration of NH_4^+ in nutrient solution used depends on the variety, and that the use of 1 mmol L^{-1} of Si in the nutrient solution was sufficient to minimize the toxic effects of NH_4^+ in cucumber (Campos et al., 2016).

5.3. Silicon and the Maintenance of Chlorophyll and Photosynthetic Activity

One of effects of excessive ammonium concentration is the decrease in plant chlorophyll content and thus, a decrease in photosynthetic activity. In addition, it has been shown that Si has the ability to increase the chlorophyll content and green color index of the leaves (Ali et al., 2013, Campos et al., 2016), and to increase the photosynthetic activity (Ali et al., 2013).

In high concentrations, Si is able to activate enzymes related to oxidative stress, acting against ROS, the main cause of chloroplast degradation and consequently, chlorophyll degradation (Cao et al., 2015). Thus, the supply of Si can delay the pigment degradation and consequently, protect the leaf area, thus allowing photosynthetic activity to be maintained. The accumulation of amorphous silica in the leaf epidermis, which is responsible for improving leaf architecture, makes the leaves more upright and capable of capturing more light.

5.4. Silicon Mitigates Reactive Oxygen Species

Several studies on abiotic stresses have reported that Si activates some of the enzymes capable of mitigating ROS. Si increases the activity of superoxide dismutase (SOD), considered the primary enzyme in the fight against oxidative stress, as well as catalase (Al-Aghabary et al., 2004; Liu et al., 2015).

This higher activity of the ROS defense system may be important in mitigating phytotoxicity of NH_4^+.

5.5. Silicon and Osmotic Regulation

Stomatal conductance and transpiration are closely linked with plant osmotic regulation. With respect to this, plants with improved osmotic regulation tend to have better development. In a study on sorghum grown hydroponically with low K supply, Chen et al. (2016) found that plants receiving Si had 49% and 94% higher stomatal conductance and transpiration respectively, than plants without Si supply. The authors reported that plants supplied with Si were able to absorb more water via the roots. Where there is an excess of NH_4^+ this effect becomes more important because of the drastic reduction of the root system, which may limit the absorption of water and nutrients.

CONCLUSION

The use of NH_4^+ in hydroponic nutrient solutions, which is beneficial to plant growth, is weighed against the potential risks of NH_4^+ toxicity and as a function of factors affecting NH_4^+ absorption, and these problems can be alleviated with the use of Si.

It is important for progress in research to identify the tolerance of hydroponic crops to NH_4^+, depending on their growth stages, as well as the role of Si in modifying this tolerance.

REFERENCES

Al-Aghabary, K., Zhu, Z. & Shi, Q. (2004). Influence of silicon supply on chlorophyll content, chlorophyll fluorescence, and antioxidative enzyme activities in tomato plants under salt stress. *Journal of Plant Nutrition, 27* (12), 2101-2115.

Albornoz, F. (2016). Crop responses to nitrogen overfertilization: A review. *Scientia Horticulturae, 205*, 79–83.

Ali, S., Farooq, M. A., Yasmeen, T., Hussain, S., Arif, M. S., Abbas, F., Bharwana, S. A. & Zhang, G. P. (2013). The influence of silicon on barley growth, photosynthesis and ultra-structure under chromium stress. *Ecotoxicology and Environmental Safety, 89*, 66-72.

Apaolaza-Hernandez, L. *(2014). Can silicon partially alleviate micronutrient deficiency in plants? A review. Planta, 240* (3), 447-458.

Ariz, I., Artola, E., Asensio, A. C., Cruchaga, S., Aparicio-Tejo, P. M. & Moran, J. F. (2011) High irradiance increases NH_4^+ tolerance in *Pisum sativum*: Higher carbon and energy availability improve ion balance but not N assimilation. *Journal of Plant Physiology, 168*, 1009–1015.

Bae, M. J., Park, Y. G. & Jeong, B. R. (2010). Effect of silicate fertilizer supplemented to the medium on rooting and subsequent growth of potted plants. *Horticulture, Environment, and Biotechnology, 51* (5), 355-359.

Balkos, K. D., Britto, D. T. & Kronzucker, H. J. (2010). Optimization of ammonium acquisition and metabolism by potassium in rice (*Oryza sativa* L. cv. IR-72), *Plant, Cell and Environment, 33*, 23–34.

Barickman, T. C. & Kopsell, D. A. (2016). Nitrogen form and ratio impact Swiss chard (*Beta vulgaris* subsp. cicla) shoot tissue carotenoid and chlorophyll concentrations. *Scientia Horticulturae, 204*, 99–105.

Barreto, R. F., Prado, R. M., Leal, A. J. F., Troleis, M. J. B., Silva Junior, G. B., Monteiro, C. C., Santos, L. C. N. & Carvalho, R. F. (2016). Mitigation of ammonium toxicity by silicon in tomato depends on the ammonium concentration. *Acta Agriculturae Scandinavica. Section B - Soil & Plant Science, 66* (6), 483-488.

Bittsánszky, A., Pilinszk, K., Gyulai, G. & Komives, T. (2015). Overcoming ammonium toxicity. *Plant Science, 231,* 184-190.

Bloom, A. J., Burger, M., Kimball, B. A. & Pinter, J. P. J. (2014). Nitrate assimilation is inhibited by elevated CO_2 in field-grown wheat. *Nature Climate Change, 4,* 477-480.

Borgognone, D., Colla, G., Rouphael, Y., Cardarelli, M., Reac, E. & Schwarz, D. (2013). Effect of nitrogen form and nutrient solution pH on growth and mineral composition of self-grafted and grafted tomatões. *Scientia Horticulturae, 149,* 61–69.

Britto, D. T. & Kronzucker, H. J. (2002). NH_4^+ toxicity in higher plants: a critical review. *Journal of Plant Physioloy, 159,* 567–584.

Bybordi, A. (2010). Influence of $NO_3^-:NH_4^+$ ratios and silicon on growth, nitrate reductase activity and fatty acid composition of canola under saline conditions. *African Journal of Agricultural Research, 5* (15), 1984-1992.

Campos, C. N. S., Prado, R. M., Caione, G., Lima Neto, A. J. & Mingotte, F. L. C. (2016). Silicon and excess ammonium and nitrate in cucumber plants. *African Journal of Agricultural Research, 11* (4), 276-283.

Campos, C. N. S., Prado, R. M., Roque, C. G., Lima Neto, A. J., Marques, L. J. P., Chaves, A. P. & Cruz, C. A. (2015). Use of silicon in mitigating ammonium toxicity in maize plants. *American Journal of Plant Sciences, 6,* 1780-1784.

Cao, B., Ma, Q., Zhao, Q., Wang, L. & Xu, K. (2015). Effects of silicon on absorbed light allocation, antioxidant enzymes and ultrastructure of chloroplasts in tomato leaves under simulated drought stress. *Scientia Horticulturae, 194,* 53-62.

Chen, D., Cao, B., Wang, S., Liu, P., Deng, X., Yin, L. & Zhang, S. (2016). Silicon moderated the K deficiency by improving the plant-water status in sorghum. *Scientific Reports, 6,* Article number: 22882.

De Kreij, C., Voogt, W. & Baas, R. (1999). *Nutrient solutions and water quality for soilless cultures.* Naaldwijk, The Netherlands: Research Station for Floriculture and Glasshouse Vegetables.

De Rijck, G. & Schrevens, E. (1999). Anion Speciation in Nutrient Solutions as a Function of pH. *Journal of Plant Nutrition, 22,* 269-279.

Epstein, E. (2001). Silicon in plants: facts vs concepts. In: Datnoff, L. E., Snyder, G. H., and Korndörfer, G. H. (Eds.), *Silicon in agriculture*, (1, pp. 1-15). The Netherlands: Elsevier Science.

Fallovo, C., Colla, G., Schreiner, M., Krumbein, A. & Schwarz, D. (2009). Effect of nitrogen form and radiation on growth and mineral concentration of two Brassica species. *Scientia Horticulturae, 123*, 170–177.

Farooq, M. A., Ali, S., Hameed, A., Ishaque, W., Mahmood, K. & Iqbal, Z. (2013). Alleviation of cadmium toxicity by silicon is related to elevated photosynthesis, antioxidant enzymes, suppressed cadmium uptake and oxidative stress in cotton. *Ecotoxicology and Environmental Safety, 96*, 242-249.

Fu, Y. M., Li, H. Y., Yu, J., Liu, H., Cao, Z. Y., Manukovsky, N. S. & Liu, H. (2017). Interaction effects of light intensity and nitrogen concentration on growth, photosynthetic characteristics and quality of lettuce (*Lactuca sativa* L. Var. youmaicai). *Scientia Horticulturae, 214*, 51–57.

Gao, Q., Wang, Y., Lu, X. & Jia, S. S. (2014). Effects of exogenous silicon on physiological characteristics of cucumber seedlings under ammonium stress. *Journal of Applied Ecology, 25* (5), 1395-1400.

Gerendás, J., Zhu, Z., Bendixen, R., Ratcliffe, R. G. & Sattelmacher, B. (1997). Physiological and biochemical processes related to ammonium toxicity in higher plants. *Journal of Plant Nutrition and Soil Science, 160* (2), 239-251.

Haghighi, M. & Pessarakli, M. (2013). Influence of silicon and nano-silicon on salinity tolerance of cherry tomatoes (*Solanum lycopersicum* L.) at early growth stage. *Scientia Horticulturae, 161*, 111-117.

Hojin, Y., Jongmyung, C., Sunwan, J. & Sukki, J. (2013). Influence of $NO_3^-:NH_4^+$ ratios in fertilizer solution on growth and yield of hot pepper (*Capsicum annuum* L.) in pot cultivation. *Korean Journal of Horticultural Science and Technology, 31*, 65-71

Kaya, C., Tuna, A. L., Sonmez, O., Ince, F. & Higgs, D. (2009). Mitigation effects of silicon on maize plants grown at high zinc. *Journal of Plant Nutrition, 32* (10), 1788-1798.

Kleiber, T., Calomme, M. & Borowiak, K. (2015). The effect of choline-stabilized orthosilicic acid on microelements and silicon concentration, photosynthesis activity and yield of tomato grown under Mn stress. *Plant Physiology and Biochemistry, 96,* 180-188.

Lanquar, V. & Frommer, W. B. (2010). Adjusting ammonium uptake via phosphorylation. *Plant Signaling and Behavior, 5,* 736–738.

Li, P., Song, A., Li, Z., Fan, F. & Liang, Y. (2012). Silicon ameliorates manganese toxicity by regulating manganese transport and antioxidant reactions in rice (*Oryza sativa* L.). *Plant and Soil, 354* (1), 407-419.

Liang, Y., Hua, H., Zhu, Y., Cheng, C. & Romheld, V. (2006). Importance of plant species and external silicon concentration to active silicon uptake and transport. *New Phytologist, 172* (1), 63-72.

Liang, Y., Sun, W., Zhu, Y. G. & Christie, P. (2007). Mechanisms of silicon mediated alleviation of abiotic stresses in higher plants: a review. *Environmental Pollution, 147* (2), 422-428.

Liang, Y., Yang, C. & Shi, H. (2001). Effects of silicon on growth and mineral composition of barley grown under toxic levels of aluminum. *Journal of Plant Nutrition, 24* (2), 229-243.

Lim, M. Y., Lee, E. J., Jana, S., Sivanesan, I. & Jeong, B. R. (2012). Effect of potassium silicate on growth and leaf epidermal characteristics of begonia and pansy grown *in vitro*. *Korean Journal of Horticultural Science & Technology, 30* (5), 579-585.

Liu, G., Du, Q. & Li, J. (2017) Interactive effects of nitrate-ammonium ratios and temperatures on growth, photosynthesis, and nitrogen metabolism of tomato seedlings. *Scientia Horticulturae, 214,* 41–50.

Liu, Y. & Wirén, N. (2017). Ammonium as a signal for physiological and morphological responses in plants. *Journal of Experimental Botany,* erx086, 1-12.

Lopes, M. S. & Araus, J. L. (2006). Nitrogen source and water regime effects on durum wheat photosynthesis and stable carbon and nitrogen isotope composition. *Physiologia Plantarum, 126* (3), 435-445.

Ma, J. F. & Yamaji, N. (2006). Silicon uptake and accumulation in higher plants. *Trends in Plant Sciences, 11* (8), 392-397.

Ma, J. F., Yamaji, N. & Mitani-Ueno, N. (2011). Transport of silicon from roots to panicles in plants. *Proceedings of the Japan Academy. Series B, Physical and biological sciences*, 87 (7), 377-385.

McKeague, J. A. & Cline, M. G. (1963). Silica in the soil. *Advances in Agronomy*, 15, 339-396.

Mitani, N. & Ma, J. F. (2005). Uptake system of silicon in different plant species. *Journal of Experimental Botany*, 56 (414), 1255-1261.

Nikolic, M. & Römheld, V. (2003). Nitrate does not result in iron inactivation in the apoplast of sunflower leaves. *Plant Physiology*, 132, 1303-1314.

Prado, R. M. (2008). *Nutrição de Plantas*. 1.ed. São Paulo: Editora UNESP.

Prado, R. M. & Esteban, A. V. (2011). Influência de formas de nitrogênio e do pH na correção da deficiência de ferro no girassol. *Revista de Ciências Agrárias*, 34, 210-217.

Reports of the Scientific Committee for Food, 38th series, Opinion of the Scientific Committee for Food on nitrates and nitrite, p. 1, http://ec.europa.eu/food/fs/sc/scf/reports/scf_reports_38.pdf.

Savvas, D., Gizas, G., Karras, G., Lydakis-Simantiris, N., Salahas, G., Papadimitriou, M. & Tsouka, N. (2007). Interactions between silicon and NaCl-salinity in s soilless culture of roses in greenhouse. *Eur. J. Horticul. Sci.*, 72, 73-79.

Salsac, L., Chaillou, S., Morot-Gaudry, J. F., Lesaint, C. & Jolivet, E. (1987). Nitrate and ammonium nutrition in plants. *Plant Physiology and Biochemistry*, 25, 805-812.

Savvas, D., Gizas, G., Karras, G., Lydakis-Simantiris, N., Salahas, G., Papadimitriou, M. & Tsouka, N. (2007). Interactions between silicon and NaCl-salinity in a soilless culture of roses in greenhouse. *European Journal og Horticultural Science*, 72, 73-79.

Silva Júnior, G. B. Relação amônio e nitrato, mitigação da toxicidade amoniacal com silício e curva de acúmulo de nutrientes em mudas de maracujazeiro. *Tese (doutorado)* - Universidade Estadual Paulista (UNESP), Faculdade de Ciências Agrárias e Veterinárias, 2015. 69p. [Ammonium nitrate ratio and ammonium toxicity mitigation of silicon and nutrient accumulation curve of passion fruit seedlings. *Thesis (doctoral)* - São Paulo State University (UNESP), School of Agricultural and Veterinarian Sciences, 2015. 69p.]

Su, S., Zhou, Y., Qin, J. G., Wang, W., Yao, W. & Song, L. (2012). Physiological responses of *Egeria densa* to high ammonium concentration and nitrogen deficiency. *Chemosphere*, *86* (5), 538-545.

Tabatabaei, S. J., Yusefi, M. & Hajiloo, J. (2008). Effects of shading and $NO_3^-:NH_4^+$ ratio on the yield, quality and N metabolism in strawberry. *Scientia Horticulturae*, *116*, 264–272.

Tréguer, P., Nelson, D. M., Vanbennekom, A. J., Demaster, D. J., Leynaert, A. & Queguiner, B. (1995). The silica balance in the world ocean - a reestimate. *Science*, *268*, 375-379.

Vega-Mas, I., Marino, D., Sánchez-Zabala, J., González-Murua, C., Estavillo, J. M. & González-Moro, M. B. (2015). CO_2 enrichment modulates ammonium nutrition in tomato adjusting carbon and nitrogen metabolism to stomatal conductance. *Plant Science*, *241*, 32–44.

Wang, B. & Shen, Q. (2011). NH_4^+-N/NO_3^--N ratios on growth and NO_3^--N remobilization in root vacuoles and cytoplasm of lettuce genotypes. *Canadian Journal Of Plant Science*, *91*, 411, 417.

Wang, C., Song, H. Z., Pei, F. W., Wei, L. & Jie, L. (2010). Effects of ammonium on the antioxidative response in *Hydrilla verticillata* (L.f.) Royle plants. *Ecotoxicology and Environmental Safety*, *73* (2), 189-19

Xing, S., Wang, J., Zhou, Y., Bloszies, S. A., Tu, C. & Hu, S. (2015) Effects of NH_4^+- N/NO_3^--N ratios on photosynthetic characteristics, dry matter yield and nitrate concentration of spinach. *Experimental Agriculture.*, *51*, 151–160.

Zaghdoud, C., Carvajal, M., Ferchichi, A. & Martínez-Ballesta, M. C. (2016). Water balance and N-metabolism in broccoli (*Brassica oleracea* L. var. Italica) plants depending on nitrogen source under salt stress and elevated CO_2. *Science of the Total Environment*, *571*, 763-771.

In: Advances in Hydroponics Research
Editor: Devin J. Webster

ISBN: 978-1-53612-131-5
© 2017 Nova Science Publishers, Inc.

Chapter 3

INTERACTIVE EFFECTS OF COPPER AND LEAD ON METAL UPTAKE AND ANTIOXIDATIVE METABOLISM OF *CENTELLA ASIATICA* UNDER EXPERIMENTAL HYDROPONIC SYSTEMS

Chee Kong Yap[1,*], *Ghim Hock Ong*[2], *Wan Hee Cheng*[2], *Rosimah Nulit*[1], *Ali Karami*[3] *and Salman Abdo Al-Shami*[4]

[1]Department of Biology, Faculty of Science, Universiti Putra Malaysia, 43400 UPM Serdang, Selangor, Malaysia
[2]Inti International University, Persiaran Perdana BBN, 71800 Nilai, Negeri Sembilan, Malaysia
[3]Laboratory of Aquatic Toxicology, Faculty of Medicine and Health Sciences, Universiti Putra Malaysia, 43400 UPM, Serdang, Selangor, Malaysia
[4]Biology Department, University College of Taymma, University of Tabuk, Tabuk, Saudi Arabia

---

[*] Corresponding author: yapckong@hotmail.com (Yap, C.K).

Abstract

Centella asiatica is a medicinal plant that has widely been used for therapeutic purposes in this region. In this study, this medicinal plant were tested upon for the effects of Pb exposure on Cu accumulation and antioxidant activities which includes the activities of ascorbate peroxidase (APX),catalase (CAT), superoxide dismutase (SOD),and guaiacol peroxidase (GPX). The plants were grown in a hydroponic system under laboratory conditions. The results showed positive antagonistic effects of a second metal (Pb) on the uptake of the other single metal (Cu) by *C. asiatica*, which suggested a competition between the uptake of these metals by the plant. All four antioxidative enzymes in the *C. asiatica* showed significant increment when exposed to different metal treatments (showing the enzyme activities increased in the order of Cu>Pb). In comparison between single and combinational effect of the metals, the degree of enzyme activities in combination metals fluctuates. The present findings indicated irregular patterns of Cu accumulation whether at lower or higher levels of single or a combination of both Cu and Pb exposures. Although the four antioxidative enzymes (CAT, GPX, APX and SOD) are potential biomarkers of toxicities of Cu and Pbexposures in *C. asiatica*, further studies are still needed.

Keywords: antioxidant enzymes, *centella asiatica*, Cu, Pb

Introduction

Copper is a micronutrient for plants, which is required in minute amounts in plant tissues (Clarkson and Hanson, 1980; Howeler, 1983; Stevenson, 1986; Broadley et al., 2007). Lead (Pb), on the other hand, is a non-essential metal as well as a cumulative and slow acting protoplasmic poison. These metals originate from various anthropogenic sources such as automobiles, industrial factories, mining activities, agricultural fertilizers and pesticides (Eick et al. 1999). Heavy metals, such as Cu and Pb, that are released into the environment tend to bind to the surface layer of the soil (de Abreu et al. 1998) and are viable for uptake of plants growing on the ground.

The accumulation of excessive heavy metals in plants might lead to deleterious effects on the plant and eventually to human as the metals flow through the food chain. Sinha et al. (2006) reported a restriction on the growth plants that were exposed to 1.0 mM of Pb. The ability of plants to uptake and accumulate metals vary not only from species to species but even from tissues to tissues. Therefore, it is essential to study crop species with low capabilities of heavy metal accumulation from contaminated soil as a move to minimize the risk of heavy metals bioaccumulation through the food chain. The plant used in the present study is *Centella asiatica*(family: Umbelliferae) which not only has a reputable status as an effective folk medicine for hundreds of years (Brinkhaus et al. 2000) but it is also listed by WHO (1999) as an important medicinal herb.

According to various studies (Foyer &Noctor 2005; Foyer et al. 2009; Parra-Lobato et al. 2009), toxic heavy metals induce the production of reactive oxygen species (ROS) and triggers the antioxidant defencemechanim of a plant to protect itself from the harmful effects of ROS (Singh et al. 2007). This defence mechanism activation involves multiple steps, which eventually produces antioxidant enzymes to protect the different compartments of the plant's cells (Foyer and Noctor, 2005; Foyer etal., 2009; Parra-Lobato et al., 2009). It is worth mentioning that the effects of one metal to another can occur in three ways namely, additively, synergistically or antagonistically (Wu et al. 1995). Ong et al. (2013) investigated the effect of Pb and Cu (separately) on the accumulation of Zn as well as the antioxidant activities of *C. asiatica*(Ong et al., 2013). Many studies also focused the toxicity effects of single metal (Mocquot et al. 1996; Devi and Prasad, 1998; Rout and Das, 2009) on the growth of the plant but there is a lack of information on the interaction between Pb and Cu on the accumulation and changes of antioxidant levels in plants.Therefore, the objective of this study was to determinate the effects of added Pb on the accumulation of Cu and on antioxidant activities in *C. asiatica*.

MATERIALS AND METHODS

The young stems of *C. asiatica* were planted in a greenhouse in Taman PertanianUniversiti (TPU), Universiti Putra Malaysia for one month, so that the roots of the young plants can grow well before being transferred into hydroponic solution. The whole experiment was carried out in the greenhouse under a light density of 2500 Lux (Ong et al., 2011). The roots of the plants were washed to remove the attached soil, soaked in eighth-strength modified Hoagland nutrients solution, and then sown for one week for acclimation (Tang et al., 2009).

After one week, the plants were exposed to a series of different Pb and Cu concentrations (Table 1). The concentrations of metal were determined based on the phytotoxicity level of the trace metals towards the plant, where Cu is more toxic as compared to Pb (Kopittke et al., 2009). Hence, lower concentration exposures of Cu (Cu1; 0.10 ppm), Cu2; 0.20 ppm), Cu3; 0.30 ppm), than Pb (Pb1; 0.20 ppm, Pb2; 0.40 ppm and Pb3; 0.60 ppm) were used. The toxicity tests were carried out for 20 days in the green house, where solutions (pH 5.8) were changed every 10 days. For each treatment, each try was planted with five plants with two replicates for each treatment (Ong et al., 2011, 2013). The whole experiment was repeated twice.

Table 1. Laboratory treatments and their abbreviations

No	Treatments	Metals added
1	C	No metals added
2	Cu1	Cu (0.10 ppm)
3	Cu2	Cu (0.20 ppm)
4	Cu3	Cu (0.30 ppm)
5	CuPb1	Cu (0.10 ppm) + Pb (0.20 ppm)
6	CuPb2	Cu (0.20 ppm) + Pb (0.40 ppm)
7	CuPb3	Cu (0.30 ppm) + Pb (0.60 ppm)

The plants were harvested later on by immersing the roots into 20mmoll^{-1} Na$_2$-EDTA for 15min to removed the metals bound on the surface of the root (Yang et al., 2002). The stems and roots were separated and washed three times with double de-ionized water (Ong et al., 2011, 2013).

For Cu and Pb accumulation analysis, the plants stems and roots were digested separately following the methods suggested by Yap et al. (2010). Digested samples were then left to be cooled down before they were diluted to 40 mL with double de-ionized water. The digested solutions were filtered into acid-washed pill box and stored until metal determinations by using an air-acetylene Perkin-Elmer™ flame atomic absorption spectrophotometer model AAnalyst 800. Prodecural blanks were analysed with the samples for calibration purposes. All data from the analysis were presented in µg/g dry weight (dw) basis.

For antioxidative enzymes, the leaves and roots were selected to investigate the changes in antioxidant levels caused by single Cu and combination of Cu and Pb exposures. The antioxidant enzyme extraction carried out by following the method as suggested by Mishra et al. (2006). The supernatants from the antioxidant enzyme extraction were used for enzyme determination.

The activity of superoxide dismutase (SOD) was determined from the inhibition of the photochemical reduction of nitrobluetetrazolium (NBT) at the absorbance value of 560nm (Beauchamp and Fridovich, 1971). For catalase (CAT) activity, it was assayed based the decreasing mixture absorbance was recorded at 240 nm for 3 min (Aebi, 1984). Guaiacol peroxidase (GPX) activity was determined from a modified version of method by Hemeda and Klein (1990) where the increasing rate of absorbance from oxidation of guaiacol was measured at 470 nm for 3 min. As for ascorbate peroxidase (APX) activity, the method of Nakano and Asada (1981) was used to determine the changes of absorbance at 290nm by using an UV-spectrophotometer.

Statistical analysis such as analysis of variance (ANOVA), SNK and Post hoc test were applied for data treatment by using the statistical software, SPSS software version 17.0 for Windows. STATISTICA version

8 software was utilized to determine the t-test of any two parameters (Zar, 1996).

RESULTS

In leaves (Table 2), increased exposure of single Cu exposure resulted in a significant ($P< 0.05$) increased levels of Cu accumulation in this leave part (as compared to control). However, there were no significant difference ($P> 0.05$) among Cu1, Cu2 and Cu3 levels exposures (ranging from 29.3 to 31.1 mg/kg dw). As for combined Pb to Cu exposure, little impact on Cu accumulation in the leaves were observed at lower concentration of metals exposure (CuPb1) but the accumulation of Cu significantly increased at higher metal exposures (CuPb2 and CuPb3).

In stems (Table 2), it is also found that the increasing levels of single Cu exposures result in significant ($P< 0.05$) increased levels of Cu accumulation in this stem part when compared to control. There is a constant increasing Cu accumulation, (but not significantly $P> 0.05$) with the increasing level of Cu exposures (Cu1: 17.3 mg/kg dw, Cu2: 20.3 mg/kg dw and Cu3: 25.6 mg/kg dw). As for combined Pb to Cu exposure, little impact on Cu accumulation were observed in two lower combined metal level exposures (CuPb1 and CuPb2) but increased in Cu accumulation for CuPb3 exposure (from 25.5 to 28.2 mg/kg dw; not significant, $P> 0.05$).

In roots (Table 2), it is found that the increased levels of single Cu exposures resulted in significant ($P< 0.05$) increased levels of Cu accumulation in this root part when compared to control. However, there is no significant difference ($P> 0.05$) among Cu1, Cu2 and Cu3 levels exposures (ranging from 38.6 to 43.7 mg/kg dry weight). At lower Pb addition to Cu (CuPb1), the Cu accumulation in stem slightly decreased (from 43.7 to 38.5 mg.kg dry weight) but slightly increased in Cu accumulation of the stems at the two higher metals exposure at CuPb2 and CuPb3.

Table 2. Mean concentrations (mg/kg dry weight) of Cu in the different treatments of single Cu and combination of Cu and Pb exposures, in leaves, stems and roots of *Centella asiatica* under experimental hydroponic conditions (N = 3)

	leaves	Stem	Roots
Control	11.8	11.2	13.6
Cu1	30.3	17.3	43.7
CuPb1	29.3	17.5	38.5
Cu2	29.6	20.3	38.6
CuPb2	50.9	20.4	43.1
Cu3	31.1	25.6	42.8
CuPb3	46.2	28.2	58.5
Pb1	12.3	12.4	16.8
Pb2	13.2	12.9	15.6
Pb3	13.9	12.6	16.0

Table 3. Mean concentrations (nmol/mg/g) of antioxidant enzymes (SOD, CAT, GPX and APX) under different treatments of single Cu and combination of Cu and Pb exposures in leaves of *Centella asiatica* under experimental hydroponic conditions (N = 2)

Treatments	CAT	GPX	APX	SOD
C	0	0	0	1.02
Cu1	1.46	0.159	0.230	1.94
CuPb1	6.22	0.195	0.123	1.08
Cu2	5.34	0.239	0.420	2.26
CuPb2	5.16	0.243	0.116	1.16
Cu3	7.40	0.764	0.700	2.13
CuPb3	10.53	0.598	0.711	1.34

Note: CAT, GPX and APX for the control (C) line were blank due to the reading having been subtracted by the particular enzymes but SOD had its own control for the calculation. For the concentrations of SOD, CAT and APX, their actual values were multiplied by 1000.

For the changes of levels of antioxidant enzymes in the leaves (Table 3), at lower metals exposure (CuPb1), the addition of Pb increased the levels of CAT and GPX but decreased the levels of APX and SOD. At CuPb2 exposure, the addition of Pb did not significantly change the levels of CAT and GPX but significantly ($P < 0.05$) decreased the levels of APX and SOD. At the highest level of metal exposure (CuPb3), the addition of Pb increased the levels of CAT and APX but decreased levels of GPX and SOD.

For the changes of levels of antioxidant enzymes in the roots (Table 4), at lower metal exposures (CuPb1), the addition of Pb increased the levels of CAT (significantly, $P < 0.05$), GPX, APX (significantly, $P < 0.05$) and SOD. However, at CuPb2 exposure, the addition of Pb significantly ($P < 0.05$) decreased the levels of CAT and APX, but increased the levels of GPX and SOD. At the highest level of metals exposure (CuPb3), the addition of Pb did not change significantly ($P > 0.05$) the levels of CAT, APX and SOD, but significantly ($P < 0.05$) decreased the level of GPX.

Table 4. Mean concentrations (nmol/mg/g) of antioxidant enzymes (SOD, CAT, GPX and APX) under different treatments of single Cu and combination of Cu and Pb exposures in roots of *Centella asiatica* under experimental hydroponic conditions (N = 2)

Treatments	CAT	GPX	APX	SOD
C	0	0	0	0.805
Cu1	2.17	2.82	6.67	0.761
CuPb1	28.36	2.94	10.06	0.836
Cu2	48.34	4.07	19.31	0.810
CuPb2	20.02	6.50	2.45	0.852
Cu3	39.56	8.05	8.14	0.929
CuPb3	40.95	5.48	9.85	0.904

Note: CAT, GPX and APX for the control (C) line were blank due to the reading having been subtracted by the particular enzymes but SOD had its own control for the calculation. For the concentrations of SOD, CAT and APX, their actual values were multiplied by 1000.

DISCUSSION

The Cu accumulation in the plant is in the order of roots>leaves>stems. This is in agreement with those reported by Ong et al., 2011 on the effects of Zn addition. The result indicated that root could be a storage site for Cu and translocation of the essential Cu to stems and leaves will occur when there is a need. According to Alloway (2008), when the essential metal was transported to the leaves, the metal will be stored in the vacuoles of the leaf's cells to reduce the accumulation of toxic metals in the cytoplasm.

Significantly increase levels Cu accumulation in leaves at higher Cu and Pb exposures (CuPb2 and CuPb3) could be attributed to the competition of Pb with Cu. Interactions between Cu and other trace elements resulting in toxicity, which will eventually cause the depletion of metal concentration leading to its nutritional deficiency. Present results showed additive effects on the Cu accumulation due to the addition of Pb exposure. However, the present finding did not show an evidence of low or high concentrations of Pb inhibiting Cu uptake rates in the leaves.

There is a slightly increase of Cu accumulation in stems at the two higher metals exposure (CuPb2 and CuPb3). The effect of Pb towards the uptake of Cu was slightly increased when high Pb (Pb3) was added to the roots. This showed that Pb and Cu had a synergetic effect on each other (Tani and Barrington, 2005) in the roots. This was supported by Sharma et al. (1999) where they had reported a strong synergistic effects between metals,at a higher concentration metals mixture, in the root-internal of plants.

The increased levels of antioxidative enzymes in leaves and roots indicated that Pb was toxic in low or high Pb exposures in combination with Cu.Påhlsson (1989) pointed out that low concentrations of metal exposures (0.1 to 0.2 ppm), such as Cu, could disrupt the metabolic processes and growth of plants. According to Fernandes and Henriques (1991) high metal level exposures could also disturb metabolic mechanisms and growth inhibition. The high level metal exposures could result in oxidative stress and thus increasing the reactive oxygen species (ROS) in the subcellular compartments (Mittler et al., 2004).

The addition of Pb increased the toxicity of the available Cu which caused higher toxicity of the metal in the plant (Luo andRimmer, 1995). High concentrations of metals in the nutrient solution would generate oxidative stress to the plant when they are within the safety range (Smirnoff, 1998). This would increase the ROS levels within the subcellular compartments of the leaves and roots of the plants (Mittler et al., 2004). Sarvajeet and Narendra (2010) stated that SOD was the most effective intracellular enzymatic antioxidant that dismutates O_2^- into H_2O_2 while APX was involved in the scavenging of H_2O_2 in water-water. According to Siedlecka and Krupa (2002), SOD will be activated as long as the level of stress was within the range of the plant's defence capacity. Upadhyay and Panda (2010) stated that the activity of the antioxidant enzyme GPX was greatly increased in the combination of Cu with Zn treatment when compared to single Zn treatment.

Present findings are in agreement with many such reported studies in the literature. For example, Chi et al. (2005) reported reduction of Cu toxicity and Cu-induced NH_4^+ accumulation by nitric oxide in rice leaves. Their findings suggested that the reduced $CuSO_4$-induced toxicity and NH_4^+accumulation by sodium nitroprusside (SNP) was plausibly attributed to its ability to scavenge active oxygen species. Based on *Sesbaniadrummondii* seedlings, Israr et al. (2006) reported that the antioxidative enzymes activities of SOD, APX and GR were similar to common antioxidants where they tend to increase to a concentration of 50 mg/L Hg and slightly decreased later on. Their findings indicated the effectiveness ofSesbania'santioxidative defense mechanisms to allow the plant to accumulate and tolerate Hg induced stress. In a more recent study, Israr et al. (2011) found significant increase of enzymatic (SOD, APX, and GR) antioxidants in the *S. drummondii*seedlings when exposed to different metal treatments. They also found that exposure to combined metals increased the enzyme activities to varying degrees in comparision to single metal exposure.

CONCLUSION

The Cu accumulation in the leaves and roots were dependent on the levels of single Cu and combination of Cu and Pb exposures. The treatment of Cu with addition Pb had reduced the accumulation of Cu in leaves but increased the uptake in roots when higher Pb was added. Therefore, present findings indicated irregular patterns of Cu accumulation whether at lower or higher levels of single or combination of Cu and Pb exposures. Although the four antioxidative enzymes (CAT, GPX, APX and SOD) are potential biomarkers of toxicities of Cu and Pbexposures in *C. asiatica*, further studies are still needed to verify the claim.

ACKNOWLEDGMENTS

The authors wish to acknowledge the partial financial support provided through the Fundamental Research Grant Scheme (FRGS), [Vote no.: 5524953], by Ministry of Higher Education, Malaysia.

REFERENCES

Aebi, H. (1984). Catalase *in vitro*. *Methods in Enzymology*, *105*, 121–126.
Alloway, B. J. (2008). *Zinc in Soils and Crop Nutrition*, 2nd Edition. International Zinc Association and International Fertilizer Industry Association, Brussels, Belgium and Paris, France.
Beauchamp, C. & Fridovich, I. (1971). Superoxide dismutase: improved assays and an assay applicable to acrylamide gels. *Analytical Biochemistry*, *44*, 276–287.
Brinkhaus, B., Lindner, M., Schuppan, D.& Hahn, E. G. (2000). Review Article: Chemical, pharmacological and clinical profile of the East Asian medical plant *Centella asiatica*. *Phytomedicine*, *7*, 427-448.
Broadley, M. R., White, P. J., Hammond, J. P., Zelko, I. & Lux, A. (2007). Zinc in plants. *New Phytologist*,*173*(4), 677.

Chi, Y. C., Tung, H. K. &Huei, K. C. (2005). Nitric oxide reduces Cu toxicity and Cu-induced NH_4^+ accumulation in rice leaves. *Journal of Plant Physiology*, *162*, 1319–1330.

Clarkson, D. T. & Hanson, J. B. (1980). The mineral nutrition of higher plants. *Annual Review of Plant Physiology*, *31*, 239-298.

De Abreu, C. A., de Abreu, M. F. & de Andrade, J. C. (1998). Distribution of lead in the soil profile evaluated by DTPA and Mehlich-3 solutions. *Bragantia*, *57*, 185-192.

Devi, S. R. & Prasad, M. N. V. (1998). Copper toxicity in *Ceratophyllumdemersum* L. (Coontail), a free floating macrophyte: Response of antioxidant enzymes and antioxidants. *Plant Science*,*138*(2), 157-165.

Eick, M. J., Peak, J. D., Brady, P. V. &Pesek, J. D. (1999). Kinetics of lead adsorption and desorption on goethite: Residence time effect. *Soil Science*, *164*,28–39.

Fernandes, J. C. &Henriques, F. S. (1991). Biochemical, Physiological, and Structural Effects of Excess Copper in Plants. *The Botanical Review*, *57*, 3.

Foyer, C. H. &Noctor, G. (2005). Redox homeostasis and antioxidant signaling: a metabolic interface between stress perception and physiological responses. *The Plant Cell*, *17*(7), 1866–1872.

Foyer, C. H., Noctor, G., Buchanan, B., Dietz, K. J. & Pfannschmidt, T. (2009). Redox regulation in photosynthetic organisms: signaling, acclimation and practical implications. *Antioxidants & Redox Signaling*, *11*(4), 861–905.

Hemeda, H. M. & Klein, B. P. (1990). Effects of naturally occurring antioxidants on peroxidase activity of vegetable extracts. *Journal of Food Science*, *55*, 184–185.

Howeler, R. H. (1983). Study of some tropical plants for the diagnosis of nutritional problems. Cali, Colombia: International Centre for Tropical Agriculture (in Spanish).

Israr, M., Sahi, S., Datta, R.& Sarkar, D. (2006). Bioaccumulation and physiological effects of mercury in Sesbaniadrummondii. *Chemosphere*, *65*, 591–598.

Israr, M., Jewell, A., Kumar, D.&Sahi, S. V. (2011). Interactive effects of lead, copper, nickel and zinc on growth, metal uptake and antioxidative metabolism of Sesbaniadrummondii. *Journal of Hazardous Materials*,*28*, *186*(2-3),1520-6.

Kopittke, P. M., Blamey, F. P. C., Asher, C. J. & Menzies, N. W. (2009). Trace metal phytotoxicity in solution culture: a review. *Journal of Experimental Botany*, *61*(4), 945-954.

Luo, Y. M. & Rimmer, D. L. (1995). Zinc-copper interaction affecting plant growth on a metal-contaminated soil. *Environmental Pollution*, *88*(1), 78-93.

Mishra, S., Srivastava, S., Tripathi, R. D., Govindarajan, R., Kuriakose, S. V. & Prasad, M. N. V. (2006). Phytochelatin synthesis and response of antioxidants during cadmium stress in BacopamonnieriL. *Plant Physiology and Biochemistry*, *44*, 25–37.

Mittler, R., Vanderauwera, S., Gollery, M. & Breusegem, F. V. (2004). Abiotic stress series. Reactive oxygen gene network of plants. *Trends in Plant Science*, *9*(10), 490– 498.

Mocquot, B., Vangronsveld, J., Clijsters, H. & Mench, M. (1996). Copper toxicity in young maize (*Zea mays* L.) plants: effects on growth, mineral and chlorophyll contents, and enzyme activities. *Plant and Soil*, *182*(2), 287-300.

Nakano, Y. & Asada, K. (1981). Hydrogen peroxide is scavenged by ascorbate-specific peroxidase in spinach chloroplasts. *Plant & Cell Physiology*, *22*, 867–880.

Ong, G. H., Yap, C. K., Marziah, M. & Tan, S. G. (2011). The effect of Cu exposure on the bioaccumulation of Zn and antioxidant activities in different parts of *Centella asiatica*. *Asian Journal of Microbiology, Biotechnology and Environmental Sciences*, *13*(3), 387–392.

Ong, G. H., Yap, C. K., Maziah, M. & Tan, S. G. (2013). Synergistic and antagonistic effects of zinc bioaccumulation with added lead and the changes in antioxidant activities in leaves and roots of *Centella asiatica*. *SainsMalaysiana*,*42*(11),1549–1555.

Påhlsson, A. M. B. (1989). Toxicity of heavy metals (Zn, Cu, Cd, Pb) to vascular plants. *Water, Air & Soil Pollution*, *47*(3-4), 287-319.

Parra-Lobato, M. C., Fernandez-Garcia, N., Olmos, E., Alvarez-Tinaut, M. C. & Gomez- Jimenez, M. C. (2009). Methyl jasmonate-induced antioxidant defence in root apoplast from sunflower seedlings. *Environmental and Experimental Botany*, *66*(1), 9–17.

Rout, G. R.& Das, P. (2009). Effect of Metal Toxicity on Plant Growth and Metabolism: I. Zinc. *Sustainable Agriculture*, *7*, 873-884.

Sarvajeet, S. G. & Narendra, T. (2010). Reactive oxygen species and antioxidant machinery in abiotic stress tolerance in crop plants. *Plant Physiology and Biochemistry*, *48*, 909-930.

Sharma, S. S., Schat, H., Vooijs, R. &VanHeerwaarden, L. M. (1999). Combination toxicology of copper, zinc, and cadmium in binary mixtures: Concentration-dependent antagonistic, nonadditive, and synergistic effects on root growth in *Silene vulgaris*. *Environmental Toxicology and Chemistry*, *18*(2), 348-355.

Siedlecka, A. & Krupa, Z. (2002). Functions of enzymes in heavy metal treated plants. In: M. N. V. Prasad & S. Kazimierz (Eds.), Physiology and bichemistry of metal toxicity and tolerance in plants., (pp. 314-317). Netherlands: Kluwer.

Singh, B. K. (2007). Studies on variability and heterosis of important economic and nutritive traits in cabbage. Ph.D. Thesis. IARI, Pusa, New Delhi, India.

Sinha, P., Dube, B. K., Srivastava, P. & Chatterjee, C. (2006). Alteration in uptake and translocation of essential nutrients in cabbage by excess lead. *Chemosphere*, *65*, 651-656.

Smirnoff, N. (1998). Plant resistance to environmental stress. *Current Opinion in Biotechnology*, *9*, 214–219.

Stevenson, F. J. (1986). Cycles of soil: Carbon, nitrogen, phosphorus, sulfur, micronutrients. New York: John Wiley & Sons Ltd.

Tang, Y. T., Qiu, R. L., Zheng, X. W., Ying, R. R., Yu, F. M. & Zhou, Z. Y. (2009). Lead, zinc, cadmium hyperaccumulation and growth stimulation in *Arabispaniculata*Franch. *Environmental and Experimental Botany*,*66*, 126-134.

Tani, F. H. & Barrington, S. (2005). Zinc and copper uptake by plants under two transpiration rates. Part I. Wheat (Triticumaestivum L.). *Environmental Pollution, 138*, 538–547.

Upadhyay, R. & Panda, S. K. (2010). Zinc reduces copper toxicity induced oxidative stress by promoting antioxidant defense in freshly grown aquatic duckweed Spirodelapolyrhiza L.*Journal of Hazardous Materials,175* (1-3), 1081-1084.

WHO, (1999). *Monographs on selected medicinal plants*, (Vol. *1*, pp. 77–85). Geneva: World Health Organization.

Wu, Y., Wang, X., Li, Y. & Ma, Y. (1995). Compound pollution of Cd, Pb, Cu, Zn and As in plant soil system and its prevention. In: Prost, R. (Ed.), *Contaminated Soils*. Paris: 3rd International Conference on the Biogeochemistry of Trace Elements, INRA.

Yang, X., Long, X. X., Ni, W. Z. & Fu, C. X. (2002). Sedum alfrediiH.: A new Zn hyperaccumulating plant first found in China. *Chinese Science Bulletin, 47*, 1634–1637.

Yap, C. K., MohdFitri, M. R., Mazyhar, Y. & Tan, S. G. (2010). Effect of Metal-contaminated soils on the accumulation of heavy metal in different parts of *Centella asiatica*: A Laboratoty Study. *SainsMalaysiana, 39*, 347-352.

Zar, J. H. (1996). *Biostatistical analysis* (3rd. edition). New Jersey: Prentice Hall.

Chapter 4

PHYTOACCUMULATION OF HEAVY METALS FROM WATER USING FLOATING PLANTS BY HYDROPONIC CULTURE

Anil Kumar Giri[1], and *Prakash Chandra Mishra[1]*

[1]Department of Environmental Science, Fakir Mohan University, Balasore, Odisha, India

ABSTRACT

Advanced technologies and technical progress water contamination by various pollutants is one of the most significant environmental problems more widespread in the future. The present aimed to develop the phytoextraction potential of the free floating aquatic plant for heavy metals from aqueous solution. The accumulation, relative growth and bio-concentration factor of metal ions at different concentrations of chromium solution significantly increased ($P<0.05$) with the passage of time.plants treated with 4.0 mg/L of chromium (VI) accumulated the highest concentration of metal in roots (1320mg/kg, dry weight) and shoots (260mg/kg, dry weight); after 15 days. To ascertain the mechanism of the process the plant biomasswas characterized by SEM-EDX and FTIR techniques. Microwave-assisted extraction efficiency is

---

* E-mail: anilchemnit@gmail.com, Tel.: +91-9439949398.

investigated by comparison of the results with after wet digestion. Chromatograms are obtained for chromium species in plant shoot biomass by using HPLC-ICP-MS. For extraction of chromium ions from plant materials using two extractant solution, 95.14% extracted by 0.02 M ethylenediaminetetraacetic acid (EDTA), and 90.24% extracted by a HCl at 80°C duration 25 minutes. Phytoextraction technique is environmental friendly, cost-effective, aesthetically pleasing, technologically feasible, long-term applicability, and ecological aspect.

Keywords: *Eichhornia crassipes*, Phytoremediation, Chromium(VI), Microwave assisted extraction, Bio-concentration factor, Chromatograms

1. INTRODUCTION

Phytoaccumulation is a novel technology that uses terrestrial and aquatic plants to degrade, extract, contain, or immobilize contaminants from soil and water. This technology effort have focused on the use of green plants to degradation of organic contaminants, usually in concert with root rhizosphere microorganisms, or remove hazardous heavy metals from soils or water (Salt et al. 1995; Licht et al. 1995). Toxic metal pollution of waters and soils is major environmental problem and most conventional remediation approaches do not provide acceptable solution. The use of especially selected and engineered metal accumulating plants for environmental clean-up is an emerging technology called Phytoremediation. Soil and water in many areas of the world are polluted with toxic trace elements. These include metalloids such as selenium (Se) and arsenic (As), as well as heavy metals, e.g., cadmium (Cd), lead (Pb), chromium (Cr) and mercury (Hg). Because of the acute toxicity of these elements, there is an urgent need to develop technologies to remove or detoxify them. This technology has been receiving attention lately as an innovative, cost-effective alternative to the more established treatment methods used at hazardous waste sites (Raskin et al. 1994; Khellof and Zerdaoui 2012). Although microorganisms have also been tested for remediation potential, plants have shown the greater ability to withstand and accumulate high concentrations of toxic metals and chemicals. Phytodetoxification involves the ability of plants to change the chemical species of the contaminant to a less toxic form, as occurs when plants take up toxic hexavalent chromium (Cr) and convert it to non toxic trivalent Cr. Some microbial based population based to capture, accumulate or breakdown contaminants as indicators of water or to reduce contaminants mobility. Algae serve as indicators of water pollution since they respond typically many ions and toxicants. Blue-green algae are ideally suitable to play a dual role of treating wastewater in the process of effective utilization of different constituents essential for growth leading to enhanced biomass production. Advantages of this techniques are Environmentally friendly, cost-effective, and aesthetically pleasing, metals absorbed by the plants may be extracted from harvested plant biomass, may reduce the entry of contaminants into the environment by preventing their leakage into the groundwater systems.It is potentially the least harmful method because it uses naturally occurring organisms and preserves the environment in a more natural state.

Some limitation of this techniques are slow growth and low biomassrequire a long-term commitment, with plant-based systems of remediation, it is not possible to completely prevent the leaching of contaminants into the groundwater, the survival of the plants is affected by the toxicity of the contaminated land and the general condition of the soil, bio-accumulation of contaminants, especially metals, into the plants which then pass into the food chain, from primary level consumers upwards or requires the safe disposal of the affected plant material.

2. VARIOUS PHYTOREMEDIATION TECHNIQUES

Phytoremediation is a low-cost technique that uses plant-associated microorganisms to degrade or decontaminate soils, groundwater and air contaminated by organic and inorganic contaminants. Several mechanisms of Phytoremediation may be involved in contaminant removal from soils or water:

2.1. Phytoextraction

Phytoextraction, also called phytoaccumulation technique utilizes the ability of certain plants to take up contaminants from the soil and water and accumulate them in their tissues, which can then be harvested and removed from the site. Hyperaccumulation are often used in this process for their enhanced capacity to extract contaminants from their surrounding environment.

2.2. Phyto-Degradation

Phyto-degradation is the metabolism of contaminants within plant tissues. Plants produce enzymes, such as dehalogenase and oxygenase that help catalyze degradation. Investigations are proceeding to determine if both aromatic and chlorinated aliphatic compounds are amenable to phyto-degradation.

2.3. Rhizodegradation

Rhizodegradation also called enhanced rhizosphere biodegradation, phytostimulation, or plant-assisted bioremediation/degradation, is the breakdown of contaminants in the soil through microbial activity that is enhanced by the presence of the rhizosphere and is a much slower process than phytodegradation. Microorganisms (yeast, fungi, or bacteria) consume and digest organic substances for nutrition and energy. Certain microorganisms can digest organic substances such as fuels or solvents that are hazardous to humans and break them down into harmless products through biodegradation.

2.4. Rhizofiltration

Rhizofiltration is the adsorption or precipitation onto plant roots or absorption into the roots of contaminants that are in solution surrounding the root zone. The plants to be used for cleanup are raised in greenhouses with their roots in water rather than in soil (Cunningham et al. 1995). Acclimate the plants once a large root system has been developed, contaminated water is collected from a waste site and brought to the plants where it is substituted for their water source.

2.5. Phytostabilization

Phytostabilization means Plants prevent contaminants from migrating by reducing runoff, surface erosion, and ground water flow rates. "Hydraulic pumping" can occur when tree roots reach ground water, take up large amounts of water, control the hydraulic gradient, and prevent lateral migration of contaminants within a ground water zone.

2.6. Phytovolatization

Phytovolatization is the uptake and transpiration of a contaminant by a plant, with release of the contaminant or a modified form of the contaminant to the atmosphere from the plant. Phytovolatilization occurs as growing trees and other plants take up water and the organic contaminants. Some of these contaminants can pass through the plants to the leaves and volatilize into the atmosphere at comparatively low concentrations.

3. GENETICS MECHANISMS INVOLVED IN PHYTOREMEDIATION TECHNIQUES

Genetic engineering are powerful methods for enhancing natural Phytoremediation capabilities, or for introducing new capabilities into plants. Genes for Phytoremediation may originate from a microorganism or may be transferred from one plant to another variety better adapted to the environmental conditions at the cleanup site. Microorganisms are very diverse, they includes bacteria, fungi, green algae and protists. The mechanisms involved in bacterial metal resistance result from either the active efflux pumping of the toxic metal out of the cell, or the enzymatic detoxification (generally redox chemistry) converting a toxic ion into a less toxic or less available metal ion.

Detoxification of metals by the formation of complexes is used by most eukaryotes. Metallothioneins (MTs) or phytochelatins (PCs) are low molecular weight (6-7 kDa), cysteine-rich proteins found in animals, higher plants, eukaryotic microorganisms and some prokaryotes (Hamer 1986).

They are divided into three different classes on the basis of their cysteine content and structure. The Cys-Cys, Cys-X-Cys and Cys-X-X-Cys motifs (in which X denotes any amino acid) are characteristic and invariant for metallothioneins (Robinson et al. 1993). The biosynthesis of PCs is induced by many metals including Cd, Hg, As, Ag, Cu, Ni, Au, Pb and Zn; however, Cd is by far the strongest inducer (Grill et al.1987). The metal binds to the constitutively expressed enzyme, γ-glutamylcysteinyl dipeptidyl transpeptidase (PC synthase), thereby activating it to catalyse the conversion of glutathione (GSH) to phytochelatin (Zenk 1996). Glutathione, the substrate for PC synthase, is synthesized from its constituent amino acids in two steps. The first step is catalysed by γ-glutamyl-Cys synthetase (γ-ECS) and the second by glutathione synthetase (GS). γ-ECS is feedback regulated by glutathione and is dependent on the availability of cysteine.

4. CHROMIUM (VI) ACCUMULATION AND REMEDIATION MECHANISM USING *EICHHORNIA CRASSIPES*

Removal of chromium (VI) was carried out using water floating macrophytes *Eichhornia crassipes* by Phytoremediation technique. High-performance liquid chromatography in conjunction with inductively coupled plasma mass spectrometry (HPLC-ICP-MS) was used to measure chromium speciation in plant material using microwave extraction processes.

4.1. Translocation of Chromium (VI)

Phytoremediation method using aquatic plants contains two uptake process i.e reversible metal binding and irreversible ion sequestration (Dutton and Fisher 2011; Mohanty and Patra 2012). The accumulation of hexavalent chromium ions by free floating plant *E. crassipes* at 25°C with different concentrations and exposure times was analyzed and were presented in Table 1. From the data in Table 1 it is clear that there was an increase in the chromium (VI) accumulation in shoots and roots when chromium (VI) concentration and exposure times were increased ($P<0.05$). Plants treated with 4.0 mg/L of Cr (VI) after day 15 accumulated the highest level of metal in shoots (260mg/kg, dry weight) and roots (1320mg/kg, dry weight).

In the presence of excessive oxygen, chromium (III) oxidizes into chromium (VI), which is highly toxic and more soluble in water than the other forms. Chromium (VI) can easily cross the cell membrane, whereas the phosphate-sulphate carrier also transports the chromite anions. Fe, S, and P are known also to compete with Cr for carrier binding (Hadad et al. 2011; Mei et al. 2002). Metal ions penetrated plants by passive process, mostly by exchange of cations which occurred in the cell wall. All heavy metals were taken up by plants through absorption, translocation and released by excretion.

Table 1. The accumulation of chromium (VI) ions at pH 6.8 in shoots and roots of *E. crassipes* at different hexavalent chromium concentration and exposure times

(Mean ± S.D.)			
$K_2Cr_2O_7$ concentration (mg/LDays)			
	3	9	15
Shoot (mg/kg)			
0.75	10.11 ± 0.04	40.44 ± 0.43	50.45 ± 0.12
1.50	50.23 ± 0.08	80 ± 1.11	101 ± 0.16
2.50	70.24± 0.13	120 ± 0.62	140 ± 0.42
4.0	90.22 ± 0.24	240 ± 0.18	260 ± 0.05
Root (mg/kg)			
0.75	200 ± 0.42	260 ± 1.01	280 ± 1.03
1.50	300± 2.05	540 ± 0.08	560 ± 2.05
2.50	500 ± 0.11	790 ± 1.04	820 ± 1.14
4	570 ± 2.15	1220 ± 3.02	1320 ± 3.12

The standard deviation has been obtained for n=3.

The metals accumulation in *E. crassipes* increases linearly with the solution concentration in the order of leaves < stems < roots (Maine et al. 2004; Keith et al. 2006).

4.2. Remediation and Toxicity Mechanisms

The toxic effects of Chromium are primarily dependent on the metal speciation, which determines its uptake, translocation and accumulation mechanism. The oxidation state of chromium strongly influences the rate of chromium uptake. Chromium (VI) can easily cross the cell membrane and the phosphate sulphate carrier transports the chromate anions. It forms a number of stable oxyacids and anions, including $HCrO_4^-$ (Hydrochromate), $Cr_2O_7^{2-}$ (dichromate), and CrO_4^{-2} (chromate). The chromate ion has a large ionic potential and tetrahedral coordination and acts both as strong acid and an oxidizing agent. The toxic properties of Cr

(VI) originate from the action of this form itself as an oxidizing agent, as well as from the formation of free radicals during the reduction of Cr (VI) to Cr (III) inside the cell. Induction and activation of superoxide dismutase (SOD) and of antioxidant catalase are some of major metal detoxification mechanisms in plants shown in Figure 1. SOD has been proposed to be important in plant stress tolerance and provide the first line of defense against the toxic effects of elevated levels of reactive oxygen species.

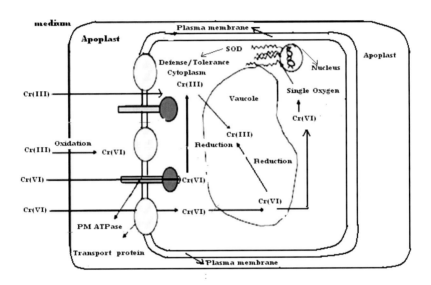

Figure 1. Model of chromium ions transport and toxicity in *E. crassipes* plant root cell.

The SODs remove $O_2^{\bullet}$ bycatalyzing its dismutation, one $O_2^{\bullet}$ being reduced to H_2O_2 and another oxidized to O_2. It has been noted that $O_2^{\bullet}$ can undergo protonation to give up - a strong oxidizing agent, $HO_2^{\bullet}$ in negatively charged membrane surfaces, which directly attack the polyunsaturated fatty acids (Halliwell 2006). Furthermore, $O_2^{\bullet}$ can also donate an electron to chromium (Cr^{6+}) to yield a reduced form of chromium (Cr^{+3}) which can then reduce H_2O_2, produced as a result of SOD led dismutation of $O_2^{\bullet}$ and $OH^{\bullet}$. The reactions through which $O_2^{\bullet}$, H_2O_2 and chromium rapidly generate $OH^{\bullet}$ which is called the Haber-Weiss reaction (Scarpeci et al. 2008).

$O_2\bullet + Cr^{6+} \longrightarrow {}^1O_2 + Cr^{+3}$

$2O_2\bullet + 2H^+ \xrightarrow{SOD} O_2 + H_2O_2\ Cr^{6+}$

$Cr^{+3} + H_2O_2 \longrightarrow Cr^{6+} + OH^- + OH\bullet$ (Fenton reaction)

5. EXTRACTION OF CHROMIUM(VI) FROM PLANT TISSUES USING DIFFERENT EXTRACTANT SOLUTION

Microwave assisted extraction as a method of sample preparation for determination of a range of chromium ions in plant samples was examined. Extraction of 260 ± 0.05 mgkg^{-1} chromium (VI) ions from *E. crassipes* shoots biomass using 0.02 M ethylenediaminetetraacetic acid (EDTA), deionized water and hydrochloric acid.

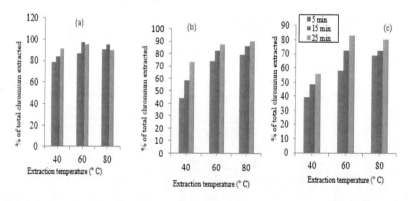

Figure 2. Total chromium extraction efficiency for *E. crassipes* shoot biomass using (a) 0.02 M EDTA, (b) HCl solution (c) deionized water at different temperature and times. Data represents the mean ± S.D (n=3).

The percentage of chromium extracted with respect to temperatures and time durations are presented in Figure 2. Chromium exists primarily in two forms, trivalent and hexavalent. Trivalent chromium is present in cationic forms as Cr^{+3} and is an essential nutrient, but hexavalent chromium is toxic and exists as an anion, either as chromate (CrO_4^{-2}) or dichromate ($Cr_2O_7^{-2}$). A common problem with chromium speciation is the well-known interconversion of Cr^{+3} and Cr^{+6}. The focus of this work is to explore chromatographic and instrumental analysis and parameters necessary to distinguish Cr^{+3} from Cr^{+6} in samples using HPLC-ICP-MS (Wolf et al. 2007; Sheehan et al. 1992).The separation is accomplished by interaction of the chromium species with the different components of the mobile phase. The Cr^{+3} forms a complex with the EDTA is retained on the column and Cr^{+6} exist in solution as dichromate. The negative charge of the chromium-EDTA complex and the negative charge of the dichromate interact with positive charge of the tetrabutylammonium hydroxide (TBAH) (Wolf et al. 2007).Therefore, the method developed in this study was tested on shoot biomass of *E. crassipes* containing 260 ± 0.05 mgkg^{-1} chromium (VI) ions. Extraction of chromium ions from plant materials using three extractant solution, 97.24% extracted by 0.02 M ethylenediaminetetraacetic acid (EDA), 72.21% extracted by double deionized water, and 87% extracted by a HCl at 60°C and time duration of 15 minutes. Chromatograms are obtained for arsenic species in plant shoot biomass by using HPLC-ICP-MS. Chromium species treated plant shoot biomass with different extractants at 60°C duration 25 minutes were shown in Figure 3.The results in figures clearly indicate that chromium species in the *E. crassipes* consisted only Cr^{+3} and Cr^{+6}, Chromium(VI) are present in maximum quantity compare with chromium (III) ions.

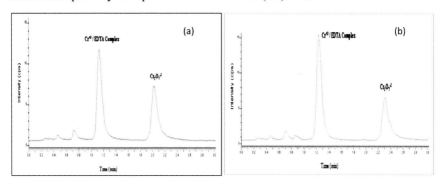

Figure 3. Chromatogram of (a) 0.02 M ethylenediaminetetraacetic acid (EDA), (b) HCl extract of freeze-dried *E. crassipes* shoot biomass from 4 mg/L chromium (VI) and duration of 25 min at 60°C treatments. Chromium species were separated on the Hamilton PRP-X100 anion-exchange column.

6. INSTRUMENTAL ANALYSIS OF AFTER REMEDIATION

The surface morphology of *E. crassipes* shoot biomass without and with removal of chromium (VI) ions during absorption process was observed with the help of SEM-EDX (JOEL model JSM-6480LV, Japan) and presented in Figure 4. It clearly reveals the surface texture and pores in the species without absorption of chromium (VI) ions. Figure 5 showsthe morphological changes with respect to shape and size of the materials after absorption of chromium (VI) ions respectively. It can be clearly observed that the surface of materials shape has changed into a new shiny bulky particles and whitish patches structure after chromium (VI) ions absorption. The EDX spectra of chromium (VI) ions unloaded was shown in Figure 4 and chromium(VI) loaded of extract materials obtained was shown in Figure 5.

Figure 4. SEM-EDX images of *E. crassipes* shoot biomass without absorption of chromium(VI) ions.

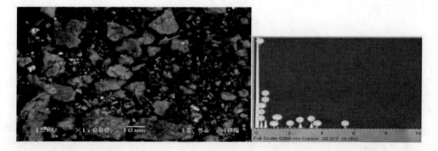

Figure 5. SEM-EDX images of *E. crassipes* shoot biomass with absorption of chromium(VI) ions.

So, it is concluded that chromium (VI) ions were adsorbed on the surface of the extract materials. These results are further confirmed with the results of FTIR spectra analysis.

Infrared spectra (Perkin Elmer Spectrum RX-I) of the *E. crassipes* shoot biomass without chromium ions loaded are obtained to determine which functional groups may have contributed to the chromium ions absorption are presented in Figure 6. There is a significant shift of few absorption peaks indicate the coordination of metal to biomass.

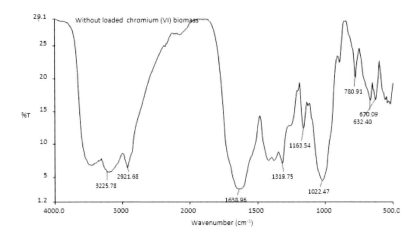

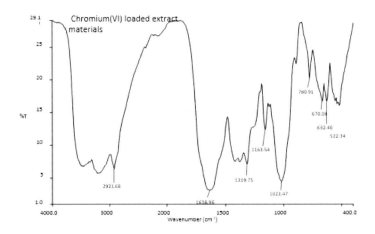

Figure 6. FTIR spectra of the *E. crassipes* shoot biomass (a) without (control) loaded chromium and (b) with chromium ions.

The band at 2918.50 cm^{-1} has been shifted insignificantly. The peaks at 1645.17 cm^{-1} have been shifted to 1638.96 cm^{-1} may be due to the complexation of carboxylic group with Cr (VI). Another shift was observed from 1418.96 cm^{-1} to 1319.75 cm^{-1}, corresponding to the complexation of nitrogen with chromium from the N-H group. Another shift was observed 1172.97 to 1163.54 cm^{-1} and 1008.50 to 1022.47 cm^{-1} may be due the interaction of nitrogen from amino group with chromium (Giri and Patel 2011). The other weak absorption peak shifted from 671.85 to 670.09 cm^{-1} and 633.04 to 632.40 corresponding to the O-C-O scissoring vibration of polysaccharide. The above changes in the spectra may be attributed to the interaction of Cr (VI) with the carboxyl, hydroxyl and amino groups present on the surface of the *E. crassipes* biomass. This clearly manifests the binding chromium to the biomass.

CONCLUSION

Nutrient culture is an efficient method for screening toxic element tolerant strains of *E. crassipes* in aqueous solution treatment. The total arsenic ions uptake is ascertained by using inductively coupled plasma mass spectroscopy (ICP-MS). The maximum concentration of chromium ions in roots (1320mg/kg, dry weight), and shoots (260mg/kg, dry weight); after 15 days. Chromium ions species speciation using HPLC-ICP-MS and extracted materials are characterized by SEM-EDX and FTIR. Least significant difference (LSD) of different concentrations of chromium solution at after 0, 3, 9 and 15 days of exposure reveals $P<0.05$. Maximum removal percentage of 97.24% extracted by 0.02 M ethylenediaminetetraacetic acid (EDA), 72.21% extracted by double deionized water, and 87% extracted by a HCl at 60°C and time duration of 15 minutes. Great removal efficiency and high chromium ions accumulation capacity make *Eichhornia crassipes* an excellent choice for Phytoremediation processes.

ACKNOWLEDGMENT

The authors are thankful to Prof. S. P. Adhikari, Vice chancellor, Fakir Mohan University, Balasore, Odisha, for necessary facilities and help in carry out the research work.

REFERENCES

Cunningham, S.D., Berti, W.R. and Hung, J. W. (1995). Phytoremediation of contaminated soil. *Trends Biotechnol.*, 13:393-397.

Dutton, J. and Fisher, N.S. (2011). Bioaccumulation of As, Cd, Cr, Hg(II), and MeHg in killifish (Fundulus heteroclitus) from amphipod and worm prey. *Sci. Total Environ.*, 409(18):3438-3447.

Giri, A.K. and Patel, R.K. (2011). Toxicity and bioaccumulation potential of Cr (VI) and Hg (II) on differential concentration by *Eichhornia crassipes* in hydroponic culture. *Water Science and Technology*, 63(5):899-907.

Grill, E., Winnacker, E.L. and Zenk, M.H.(1987), Phytochelatins, a class of heavy-metal-binding peptides from plants, is functionally analogous to metallothioneins. *Proc.Natl. Acad. Sci. U. S. A.*, 84:439-443.

Hadad, H.R., Maine, M.A., Mufarrege, M.M., Del Sastre, M.V. and Di Luca, G.A. (2011). Bioaccumulation kinetics and toxic effects of Cr, Ni and Zn on *Eichhornia crassipes*. *J. Hazard. Mater.*, 190(1-3):1016-1022.

Halliwell, B. (2006). Reactive species and antioxidants. redox biology is a fundamental theme of aerobic life. *Plant Physiol.*, 141:312-322.

Hamer, D.H., (1986). Metallothionein. *Annu. Rev.Biochem.*,55:913-951.

Keith, C., Borazjani, H.V., Diehl, S.V., Su, Y. and Baldwin, B.S. (2006). Removal of Copper, Chromium, and Arsenic by Water Hyacinths. *36th Annual Mississippi Water Resources Conference.*

Khellof, N.and Zerdaoui, M. (2012). Development of a kinetic model for the removal of zinc using the aquatic macrophyte, *Lemna gibba* L. *Water science and technology,* 66 (5): 953-957.

Licht, L.A., McCutcheon, S.C., Wolfe, N.L. and Carreira, L.H. (1995). Phytoremediation of organic and nutrient contaminants. *Environmental Science and Technology*, 29:318-323.

Maine, M.A., Sune, N.L. and Lagger, S.C. (2004). Chromium bioaccumulation: comparison of the capacity of two floating aquatic macrophytes. *Water Research*, 38: 1494-1501.

Mei, B., Puryear, I.D. and Newton R.J. (2002).Assessment of Cr tolerance and accumulation in selected plant species. *Plant and Soil*, 247: 223-231.

Mohanty, M. and Patra, H. K. (2012). Phytoremediation potential of paragrass-An in situ approach for chromium contaminated soil. *Int. J. of Phytoremediation*, 806-819.

Raskin, I., Nanda-Kumar, P.B.A., Dushenkov, S., Salt, D.E. and Ensley, B.D. (1994). Removal of radionuclides and heavy metals from water and soil by plants. *OECD Document, Bioremediation*, 345-354.

Robinson, N.J., Tommey, A.M., Kuske, C. and Jackson, P.J. (1993). Plant metallothioneins. *Biochem. J.,* 295:1-10.

Salt, D.E., Blaylock, M., Kumar, N.P.B.A., Dushenkov, V., Ensley, B.D., Chet, I. and Raskin. I. (1995). Phytoremediation - A novel strategy for the removal of toxic metals from the environment using plants. *Nature Biotechnology*,13(5):468-474.

Scarpeci, T.E., Zanor, M.I., Carrillo, N., Mueller-Roeber, B. and Valle, E. M. (2008). Generation of superoxide anion in chloroplasts of Arabidopsis thaliana during active photosynthesis: a focus on rapidly induced genes. *Plant Mol. Biol.*, 66:361-378.

Sheehan, P., Ricks, R., Ripple, S. and Paustenbach, D. (1992). Field evaluation of a sampling and analytical method for environmental levels of airborne hexavalent chromium. *Am. Ind. Hyg. Assoc. J.*, 53(1):57-68.

Wolf, R. E., Morrision, J.M. and Goldhaber, M.B. (2007). Simultaneous determination of Cr(III) and Cr(VI) using reversed-phased ion-pairing liquid chromatography with dynamic reaction cell inductively coupled plasma mass spectrometry. *Journal of Analytical Atomic Spectrometry*, 22:1051-1060.

Zenk, M.H. (1996). Heavy metal detoxification in higher plants - a review. *Gene,* 179:21-30.

In: Advances in Hydroponics Research
Editor: Devin J. Webster
ISBN: 978-1-53612-131-5
© 2017 Nova Science Publishers, Inc.

Chapter 5

DEVELOPMENT OF NUTRITIVE SOLUTIONS FOR HYDROPONICS USING WASTEWATER

André Luís Lopes da Silva[1,*],
Micheli Angelica Horbach[2], *Gilvano Ebling Brondani*[3]
and Carlos Ricardo Soccol[1]

[1]Bioprocess Engineering and Biotechnology Department,
Federal University of Paraná, Curitiba, Brazil
[2]Department of Agricultural Sciences, State University of Western Paraná, Marechal Cândido Rondon, Brazil
[3]Department of Forest Sciences, Federal University of Lavras, Lavras, Brazil

ABSTRACT

Hydroponics is a technique of growing plants without soil, in water containing dissolved nutrients. Nowadays, the hydroponics has several applications, such as: (1) Aid in the acclimatization of micropropagated plants, (2) Treatment of wastewaters, (3) Method to produce foods in places of the earth or outside that cannot support crops in the soil, (4) Sustaining the growth of ministumps or microstumps to produce cuttings

---

[*] Corresponding Author Email: clonageinvitro@yahoo.com.br.

for plant propagation, (5) Allowing to collect roots without kill the plant to obtain active principles present in the roots in the case of medicinal plants and (6) allowing crop production in larger amount and quality. However, the practice of hydroponics is expensive and one alternative to reduce the production costs is to develop nutritive solutions from by-products. Several wastewaters that can be used to develop these nutritive solutions are the vinasse, cassava wastewater, corn steep liquor and domestic effluent. Mineral nutrients must be determined to establish the chemical supplementation in some cases. Sometimes, decantation and filtration of wastewaters must be carried out to avoid toxicity to the crops. Adequate dilution of the wastewater is important to avoid salt stress (e.g., only 10% vinasse was used to develop a vinasse nutritive solution for lettuce, rocket and watercress). Crop productivity using wastewater solutions is similar to the commercial solutions used for hydroponics. The use of wastewaters to formulate nutritive solutions represents a rational alternative to wastewaters disposal and adds value to what is currently considered a waste product.

Keywords: soilless cultivation, alternative fertilizer solution, by-products, industrial residue, effluent

INTRODUCTION

Hydroponic is a cultivation system of growing plants without soil and in water containing dissolved nutrients. These hydroponics solutions have the essential nutrients required for plant growth and development. Soilless cultivation system can be utilized in a wide range of different situations and climates, since plant production in arid or cold climates and up to the acclimatization of micropropagated plants originated from tissue culture. Furthermore, soilless cultivation was successfully applied for the production of vegetative propagules from ministumps or microstumps of several species and for essential oil and bioactive molecules production, present in roots of medicinal or aromatic plants.

Hydroponic techniques enable the optimal control of plants physiological quality during cultivation process and higher yield compared to soil cultivation. For example, the acclimatization of micropropagated plants to *ex vitro* conditions by hydroponic techniques is a form to reduce moisture loss and mortality during acclimatization. Additionally, such cultivation system can expand plant production in small areas.

Hydroponics represents also an excellent tool to evaluate different abiotic stresses, such as salinity and toxicity of heavy metals, besides, it allows to evaluate the effects of the deficiency of certain nutrients for the simple reason that it has a great control of the nutritional conditions (Silva et al., 2012).

Among the various classifications of hydroponic systems are open and closed systems. In open system, the nutritive solution is not recycled, while the stock solution is recycled in closed systems. This is important when we consider the use and reuse of water by crops, since, in case of closed hydroponics systems, the water losses are lower than in soil cultivation.

Therefore, an alternative to cultivate plants in hydroponics is the use of wastewaters, or recycle waters. Industrial by-products are a major environmental concern and represents an elevated cost to treat and disposal. Wastewaters applicability in hydroponics is a cost effective and environmental solution to treat and reuse water and nutrients from industrial waste. In this way, wastewaters can be an inexpensive source of nutrients and minerals for cultivation in a hydroponic system. This alternative to hydroponic solutions is especially interesting to prevent pollution with the discard of wastewaters in rivers or streams and, to reduce the waste in cultivation systems.

Alternative Uses of Wastewaters in Hydroponic Solution

Sugarcane-based ethanol production represents a key cultivation to the biofuels industry. For the 2015-16 crop year, sugarcane production is estimated at 658.7 million tons in Brazil, an increase of 3.8% in relation to the previous harvest. Likewise, total ethanol production is expected to exceed 29 billion liters (Conab, 2015). One of the problems facing ethanol production is that the activity generates liters of vinasse. Vinasse is a by-product of alcohol production, containing pollutant substances, and can be an environmental problem if not correctly treated and dispose. In this context, an average proportion of 12-15 L is generated for each liter of alcohol produced (Santos et al., 2013), more than enough to grow into a problem for treatment and final discard.

Cassava wastewater effluent, a wastewater from cassava flour or starch production is a residue of unpleasant odor, that contributes to the pollution of water resources with the production of large amount of effluent rich in organic matter. Moreover, certain products such as manioc starch, used large amounts of water and generate dust (Howeler et al., 2001).

Others industry by-products that can be also a challenge include corn steep liquor, a residue produced in large amount in corn processing industry and waste from potato processing industry. All these residues could be highly polluting if not treated and properly discarded.

Wastewaters are usually rich in nutrients and can be utilized in several ways. Among the alternative forms of wastewaters disposal and reuse are the utilization in biogas production, as a fertilizer in plant cultivation, as a culture medium for plant tissue culture, in bioremediation process and to produce biopolymers.

Mineral Composition

Organic and inorganic elements present in wastewaters are uncertain and varies according to its source, culture, region of production, industrial process, among others. Thereby, some treatments to convert the wastewater in a useful product may be required, depending on the final use of this residue. Additionally, the chemical analysis of nutrients must be determined, as well as the suitable level of the wastewater (i.e., dilution) to be used in the cultures. The suitable level must be determined using different dilutions as treatments on crop cultivation.

The high concentration of organic matter present in wastewaters results in an increased in consumption of dissolved oxygen and, may not be safer to use as hydroponic solution. Biochemical oxygen demand (BOD) and chemical oxygen demand (COD) are one of the basics characteristics to take care when working with by-products, that usually presents high levels of BOD and COD. For example, sugarcane vinasse has BOD between 20,000 and 35,000 mg L^{-1}, meaning that this waste is extremely harmful to the flora and fauna of lakes and rivers (Silva et al., 2007).

On the other hand, some treatments can decrease BOD and COD. After a process of decantation and filtration, the treated sugarcane vinasse presents a BOD of 6,524.6 mg L^{-1} (Silva et al., 2014). In cassava wastewater, BOD was reduced with the employed of anaerobic biodigestion along with the bonus of energy production (biogas) (Barana and Cereda, 2000).

Other alternative to detoxify wastewater with high BOD is the previous culture of algae. An example of this proccess can be demonstrated using the microalga (*Haematococcus pluvialis*), whereas this brief culture (15 days) resulted in a lower phytotoxicity and phenolic compounds, with higher levels of SO_4, Na, Cl and Ca (Gollo et al., 2016). Besides, the removal of nitrogen, phosphorus and COD from cassava wastewater can be carried out with nonliving macrophyte *Salvinia*, with 98% of the nitrogen removed after 24 hours (Hasan et al., 2015).

Furthermore, it is necessary to know the dilution effect of this residue on plants, as a way to decrease toxicity and resulting in a safer solution. The step of decanting is a crucial phase in order to remove the excess of solids and organic material, while filtration decreases fragments that are not been extracted before. This process was described by Santos et al., (2013): for decantation, 40 g of $Ca(OH)_2$ and 30 g of $Al_2(SO_4)_3$ was added to 100 L of vinasse for 24 hours; after decantation, the vinasse was submitted to filtration procedure, that consist of a PVC tube of 20 cm x 1.80 m with a filter with six layers (large and small rock fragments, large and fine particles of sand and two layers of a non-permeable membrane with a metallic drain); the final result is a clear product, from which about 90% of the solids were removed and the pH was elevated from 3.5 to 6.2.

In the development of hydroponic nutritive solution from wastewaters, the nutrients present in the solution may vary according to its source and some kind of supplementation can be necessary (Table 1). A sugarcane vinasse hydroponics solution was established using 10% treated sugarcane vinasse (decanted and filtered) and there was supplemented with 750 mg L^-

^{1} Ca(NO$_3$)$_2$.6H$_2$O, 500 mg L^{-1} KNO$_3$, 150 mg L^{-1} NH$_4$PO$_4$, 400 mg L^{-1} MgSO$_4$.7H$_2$O as well as 1 mL L^{-1} micronutrient stock solution (2.34 mg L^{-1} MnCl$_2$.H$_2$O, 0.88 mg L^{-1} ZnSO$_4$.7H$_2$O, 0.2 mg L^{-1} CuSO$_4$.5H$_2$O, 2.04 mg L^{-1} H$_3$BO$_3$ and 0.26 mg L^{-1} Na$_2$MoO$_4$.2H$_2$O) (Table 1) (Santos et al., 2013).

Table 1. Composition of different vinasse sources

Wastewater	Composition (mg L^{-1})	Citation
Vinasse	K$_2$O (1,200); CaO (180); MgO (88); pH 4.3	Paula et al., (1999)
Vinasse	Ca (1,734.67); Cu (86); Fe (17); Mn (14); Al (0.01) SO$_4$ (820); PO$_4$ (78); NO$_2$ (600); pH 4.3	Ahmed et al., (2013)
Pure vinasse	Cl (59.4); SO$_4$ (1,680); Na (8.6); K (1,620); Ca (3,160); Mg (162.4); PO$_4$ (560); Fe (44.9); Mn (4.9); Zn (1.2)	Santos et al., (2013)
Decanted and filtered vinasse	Cl (37.1); SO$_4$ (1,458); Na (7); K (1,760); Ca (1,642); Mg (101.8); PO$_4$ (380); NO$_3$ (0.66); NH$_4$ (47.6); Fe (27.9); Mn (2.88); Zn (0.75)	Santos et al., (2013)

The Efficiency of Alternative Hydroponic Solutions

Hydroponics solution composed with wastewater can be adapted according to the needs of each crop culture. Alongside, the use of wastewaters to formulate alternative nutritive solution to hydroponic systems is especially interesting to prevent pollution and the disposal of organic materials in water courses. Finally, wastewaters are an inexpensive source of nutrients and minerals from abundant and readily available materials, with the addition of industrial waste reuse.

The hydroponics solution performed with sugarcane vinasse (Santos et al., 2013) was extremely efficient to cultivate lettuce, rocket and watercress, it was compared with other commercial solution and the result was the same and in some traits, superior.

Corn steep liquor can be a beneficial fertilizer for use in hydroponic systems because, in addition to providing nutrients, this by-product acts as a biostimulant and pathogens inhibitor, promoting the beneficial microorganisms development (Chinta et al., 2014).

Another form of recycle water is the use of tilapia residual water for hydroponic production of lettuce, with enough quality even without the use of commercial fertilizers (Geisenhoff et al., 2016).

CONCLUSION

By-products of alcohol production as vinasse or another wastewater (cassava wastewater, corn steep liquor, domestic effluent and others) could be an excellent substitute of hydroponic solutions performed of analytical degree reagents. The development of nutrient solutions from wastewaters is not only a form to reduce the effective cost of plant cultivation, but it is a solution for a much larger problem, that is the correct dispose of the waste generated by industrial process.

REFERENCES

Ahmed, O., Sulieman, A. M. E., Elhardallou, S. B. 2013. Physicochemical, chemical and microbiological characteristics of vinasse, a by-product from ethanol industry. *Am. J. Biochem.* 3(3), 80-83.

Barana, A. C., Cereda, M. P. 2000. Cassava wastewater (cassava wastewater) treatment using a two-phase anaerobic biodigestor. *Ciênc. Tecnol. Aliment.* 20(2), 183–86.

Chinta, Y. D., Kano, K., Widiastuti, A., Fukahori, M., Kawasaki, S., Eguchi, Y., Misu, H., Odani, H., Zhou, S., Narisawa, K., Fujiwara, K., Shinohara, M., Sato, T. 2014. Effect of corn steep liquor on lettuce root rot (*Fusarium oxysporum* f.sp. *lactucae*) in hydroponic cultures. *J. Sci. Food Agric.* 94, 2317–2323.

Conab, Companhia Nacional de Abastecimento. *Acompanhamento da Safra Brasileira de Cana de açúcar - Safra 2015/16.* v. 2. Brasília:

Conab; 2015. doi:2318-7921. [Follow-up of the Brazilian Sugarcane Harvest - Crop year 2015/2016. v. 2. Brasília: Conab].

Geisenhoff, L. O., Jordan, R. A., Santos, R. C., Oliveira, F. C., Gomes, E. P. 2016. Effect of different substrates in aquaponic lettuce production associated with intensive tilapia farming with water recirculation systems. *Eng. Agríc.* 36(2), 291-299.

Gollo, A. L., Da Silva, A. L. L, Khémeli, K., De Lima, D., Costa, J. L., et al., 2016. Developing a plant culture medium composed of vinasse originating from *Haematococcus pluvialis* culture. *Pak. J. Bot.* 48 (1), 295–303.

Hasan, S. D. M., Limons, R. S., Da Silva, F. M., Klen, M. R. F. 2015. Nonliving macrophyte *Salvinia* sp. Application for nutrient removal in starchy wastewater treatment of cassava industry. *Desalination Water Treat.* 54 (11), 3003-3010.

Howeler, R. H., Oates, C. G., Allem, A. C., Chuzel, G. H., Guy, H., Clair, H., Müller-Sämann, K. M., Okogun, Sanginga, N., Souza, L. S., Sriroth, K., Westby, A. 2001. Strategic environmental assessment: An assessment of the impact of cassava production and processing on the environment and biodiversity. *In:* Validation Forum on the Global Cassava Development Strategy. Proceedings. Food and Agriculture Organization of the United Nations (FAO); International Fund for Agricultural Development (IFAD), Rome, IT. v. 5, p. 1-137.

Paula, M. B., Holanda, F. S. R., Mesquita, H. A., Carvalho, V. D. 1999. Uso da vinhaça no abacaxizeiro em solo de baixo potencial de produção. *Pesq. agropec. bras.* 34(7), 1217-1222. [Stillage application for pineapple in soil with low potential of yield. *Pesq. Agropec. Bras.* 34(7), 1217-1222].

Santos, J. D., Da Silva, A. L. L., Costa, J. L., Scheidt, G. N., Novak, A. C., Sydney, E.B., Soccol, C.R. 2013. Development of a vinasse nutritive solution for hydroponics. *J. Environ. Manage.* 15(114), 8-12. doi: 10.1016/j.jenvman.2012.10.045.

Silva, A. L. L., De Oliveira, Y., Dibax, R., Costa, J. L., Scheidt, G. N., Machado, M. P., Guerra, E. P., Brondani, G. E., Alves, S. A. O. 2012.

Hydroponics growth of *Eucalyptus saligna* Sm. on salt-stress mediated by sodium chloride. *J. Biotec. Biodivers.* 3(4), 213–218.

Silva, A. L. L., Costa, J. L., Gollo, A. L., Dos Santos, J. D., Forneck, H. R., Biasi, L. A., Soccol, V. T., De Carvalho, J. C., Soccol, C. R. 2014. Development of a vinasse culture medium for plant tissue culture. *Pak. J. Bot.* 46(6), 2195–2202.

Silva, M. A. S., Griebeler, N. P., Borges, L. C. 2007. Uso de vinhaça e impactos nas propriedades do solo e lençol freático. *Rev. Bras. Eng. Agríc. Ambient.* 11(1), 108-114. [Use of stillage and its impact on soil properties and groundwater. *Rev. Bras. Eng. Agríc. Ambient.* 11(1), 108-114].

In: Advances in Hydroponics Research　　ISBN: 978-1-53612-131-5
Editor: Devin J. Webster　　© 2017 Nova Science Publishers, Inc.

Chapter 6

HYDROPONICS FOR FEASIBILITY TEST OF BIODEGRADED FISHERY WASTE/WASTEWATER AS BIOFERTILIZER

Joong Kyun Kim[1,*], *Hyun Yi Jung*[1], *Ja Young Cho*[1] *and Geon Lee*[2]
[1]Department of Biotechnology and Bioengineering,
Pukyong National University, Busan, Korea
[2]Department of Environmental Engineeirng,
Dong-A University, Busan, Korea

ABSTRACT

This chapter provides a brief review of hydroponics used in fishery waste reuse. First, the advantages and disadvantages of hydroponics and the application of lab-scale hydroponics to the fertilizing ability test of biodegraded fishery waste are discussed. Finally, an advanced building-type hydroponics model is proposed with discussion of its characteristics and possible problems in operation.

---

[*] Corresponding author address: 45 Yongso-Ro, Nam-Gu, Department of Biotechnology and Bioengineering, Pukyong National University, Busan 608-737, Republic of Korea. E-mail: junekim@pknu.ac.kr.

Recently, the paradigm for waste policy has shifted to highly efficient social-based resource recycling, moving toward zero-emission waste management. Fishery waste is biodegradable, and the resultant culture broth is reused as biofertilizer without further wastewater treatment. The success of this process is mainly dependent on the fertilizing ability of the culture broth. To assess the feasibility of the culture broth as biofertilizer, hydroponics is a useful tool. Case studies of lab-scale hydroponics using the culture broth support hydroponics. Based on these results, an advanced building-type hydroponics model is proposed and discussed for practical use. Eventually, the practical use of the proposed hydroponic model is expected to result in development of modern hydroponics technologies.

Keywords: hydroponics, zero-emission waste management, fishery waste, biofertilizer, building-type hydroponics

1. HYDROPONIC CULTURE

1.1. Backgrounds

The human population is increasing, and is predicted to expand to 9.5 billion people within the next 40 years with an approximate increment of 36% (Bellona Foundation, 2009). This estimate undoubtedly requires security of more foodstuffs, and such demand can be offset by an increase in food production (Bellona Foundation, 2009). However, this matter is not easy when considering current agricultural systems with water security. To make matters worse, the impact of harvesting on the environment will intensify as the population increases, influencing climate change, habitat fragmentation, biodiversity loss and so on (Charles and Godfray, 2011).

Soil often brings serious limitations for plant growth, although it provides anchorage, nutrients, air, water, and so on for plant growth (Ellis et al., 1974). Such limitations originate from the presence of pathogenic organisms and nematodes, unsuitable soil reactions, unfavorable soil compaction, poor drainage, and deterioration due to erosion (Beibel, 1960). Serious problems can be found in open field agriculture where large spaces, large volumes of water, and intermittent labor are required (Butler and Oebker, 2006). To make matters worse, soil is less available in most urban and industrial areas, and fertile, cultivable and arable lands are scarce in some areas due to unfavorable geographical or topographical conditions (Beibel, 1960). Hydroculture, as a means of plant cultivation without soil, uses mineral nutrient solutions in a water solvent (Santos et al., 2013).

Hydroponics is a subset of hydroculture, and hydroculture is globally used to cultivate flowers, foliage, bedding and vegetable crops where plants use nutrients in solution for root growth (Song et al., 2004). The simplest and oldest form was a vessel of water in which inorganic chemicals were dissolved. Since then, modified methods have been developed with improved retention of nutrients and water using sphagnum peat, vermiculite, or bark chips. As a growing medium, straw bales are used in England and Canada and Rockwool (porous stone fiber) is used in Europe (Hussain et al., 2014). With development of related technologies, the construction of a series of hydroponic agricultural facilities is proposed to restrict biodiversity loss (Sam Bowring, Mission, 2015).

Modern cultivation systems, such as aeroponics, hydroponics and aquaponics systems, have some advantages in the reduction of water loss (by efficient water use) and increase of crop productivity per unit area, compared with conventional agriculture (AlShrouf, 2017). Although all three systems are very similar, hydroponics and aeroponics require additional liquid fertilizer without fish breeding in the system. However, in aquaponics, a symbiotic life of plants and fish is possible by mutual assistance: fish feed the plants and the plants clean the fish environment. Aeroponics and aquaponics are set up, based on hydroponics. To create a semi-enclosed or fully-enclosed environment in aeroponics, continued maintenance with careful attention are required. Although these techniques are water efficient, a large amount of freshwater is still consumed for the preparation of nutrient solutions essential for plant growth (Chekli et al., 2017). This can be a critical problem in most arid regions and therefore, advanced technologies to solve this problem for water and food security must be developed (Qadir et al., 2007). For irrigation of plants and crops, biologically treated wastewater has globally been a viable alternative water source (Angelakis et al., 1999). However, treated wastewater is generally not applicable due to the presence of pathogens and pollutants that are detrimental to both plants and human health [J58, J59]. Advanced treatment processes, such as ultrafiltration and reverse osmosis, are indispensable to fully eliminate any health risks (Ahluwalia and Goyal, 2007; Ferro et al., 2015).

1.2. Features

1.2.1. Established Hydroculture

1.2.1.1. Circulating Methods
In this closed system method, nutrient solution is pumped through the plant root system, and surplus solution is collected, replenished and reused. Cultivation using this method results in excellent results with diverse floras including tomato, strawberry, lettuce and lily (Hussain et al., 2014).

1.2.1.2. Nutrient Film Technique
In this technique, nutrient solution is continuously pumped into a small plastic basket, flows over the roots of the plants, and then drains back into the reservoir. Without use of growing medium, the plant roots are directly exposed to adequate supplies of water, oxygen and nutrients. However, plant roots are very susceptible to drying due to power outage and pump failure (Hussain et al., 2014).

1.2.1.3. Deep Flow Technique
In this technique, nutrient solution (2-3 cm deep) flows through sloped PVC pipes (10 cm diameter) to which plastic net pots with plants are fitted (Hussain et al., 2014). Planting material, such as old coir dust or carbonized rice husk, fill the net pots, and a small piece of net is used to prevent the planting material falling into the nutrient solution. Plants are fixed to the holes in the PVC pipes, so their roots touch the nutrient solution. The recycled solution collected in the stock tank is aerated.

1.2.1.4. Non-Circulating Methods

The nutrient solution in this method is not circulated but used only once. It is replaced when the nutrient concentration decreases or pH and electrical conductivity changes (Hussain et al., 2014).

i) Root Dipping Technique

   Plants are grown in small pots filled with a small amount of growing medium. The bottom (2–3 cm) of the pots are submerged in the nutrient solution, allowing the roots to absorb nutrients and air (Hussain et al., 2014). This method is inexpensive to set up and requires little maintenance.

ii) Floating Technique

   Plants established in small pots are fixed to a Styrofoam sheet or other light plate, which allows the pot to float on the nutrient solution filled in the container (10 cm deep). The nutrient solution is artificially aerated (Hussain et al., 2014).

iii) Capillary Action Technique

   Planting pots in different sizes and shapes with holes at the bottom are placed in shallow containers filled with the nutrient solution. The pots are filled with an inert medium in which seedlings/seeds are planted, so old coir dust mixed with sand or gravel can be used. In this technique, aeration is very important, and nutrient solution can reach inert medium by capillary action. This technique is appropriate for ornamentals, flowers and indoor plants (Hussain et al., 2014).

iv) Aeroponics

   Aeroponic systems are divided into three types (Home hydro systems, 2015). Low pressure aeroponic systems are low cost and require no special equipment compared with other types of hydroponic systems. High pressure aeroponic systems provide the roots with a fine mist (with a very small water droplet size) to increase oxygen availability; this method can be complicated and expensive. Ultrasonic fogger systems create a mist to cover the roots with a constant mist. To prevent plate clogging, the ultrasonic fogger system is often combined with the low pressure aeroponic system.

v) Aquaponics

This system provides a place in which fish and plants are grown together. The fish waste (feces) provides an organic nutrient source for the growing plants, while plants serve as a natural filter for the fish (John Facinor, 2016). Microbes including nitrifying bacteria and composting red worms are other participants that proliferate in the growing media (Gina Cavaliero, 2011).

1.2.2. Advantages and Disadvantages

1.2.2.1. Advantages

Hydroponics are helping with the challenges of climate change (Butler and Oebker, 2006). Additionally, hydroponics aid production system management for efficient utilization of natural resources and alleviation of malnutrition, resulting in superior quality, high yield, rapid harvest, and high nutrient content (Sardare and Admane, 2013). Therefore, this system can be applied to popular local crops and comply with food safety standards, and thus can produce crops at a reasonable price (Pual, 2000). In addition, plants can be stacked vertically, horizontally or in many configurations, making the most efficient use of available space. Therefore, hydroponics are effective for nations with a shortage of arable or fertile land for agriculture (Sonneveld, 2000). Furthermore, this system can provide plants with the appropriate culture conditions by managing fluctuations in influential factors such as temperature, wind, water, sunlight, etc. (Modularhydro, 2011). This culture system allows control over soil-borne diseases and pests, therefore pesticide use can be considerably reduced. Accordingly, hydroponics provide an eco-friendly environment, and reduce the cost and time taken for various tasks (Hussain et al., 2014). Table 1 represents the advantages from use of hydroponic culture (Ghehsareh et al., 2011; Olivia's growing solution, 2013; Os et al., 2002).

Table 1. Advantages of hydroponic culture

Advantage	Relevant facts
No weather dependence	Crop culture all year round
Availability of culture space	Plant culture in anywhere No issues regarding scarcity of arable or fertile land
Water saving	Only 1/20 of that required for soil-based culture
Eco-friendly culture (No use of chemicals)	Control of soil-borne diseases or pests
Less space required	20% less space than soil-based culture due to more compact growth of small plant roots
High crop yield	Almost double production yield due to faster growth under controlled environment including nutrient balance
Quality crop production	Production of healthier plants with improved nutritional value
No environment degradation	No run-off of inorganics into water systems Water conservation through reuse
No groundwater contamination	No permeation of fertilizer residues into groundwater
Less labor	Easy management with minimal maintenance
Effective marketing	Multiple, easy harvests

1.2.2.2. Disadvantages

Although hydroponics have many advantages, there are some disadvantages. When applied on a commercial scale, technical knowledge and high initial capital investment are required. Further investment is required if combined with controlled environment agriculture (Sonneveld, 2000). Great care is required with respect to plant health control, and thus high levels of management skill is required for preparation of nutrient solution, energy input for operation, maintenance of pH, DO and electrical conductivity, judgment and adjustment of nutrient deficiency, etc. (Os et al., 2002). Nutrient deficiencies or toxic symptoms can occur when pH in the nutrient solution exceeds the recommended range for a target crop. Higher electrical conductivity prevents nutrient uptake due to osmotic pressure, while lower electrical conductivity affects plant health and yield (Hussain et al., 2014). Therefore, success of this culture system (high yield of healthy plants) is dependent upon good management of these factors (Hussain et al., 2014). Risk of disease infection is low in an "open system" even in the presence of pathogens. However, this risk increases in a "closed systems" or when excess drain water flows along the roots (Olympios, 1999). Hydroponic systems are vulnerable to disease, which is almost non-existent in aquaponics (Cavaliero, 2011). For financial reasons, this system is restricted to high-value crops per area of cultivation (Hussain et al., 2014). Table 2 represents the disadvantages of hydroponic culture (Black, 2016; Diez, 2015; Hussain et al., 2014; Olivia's growing solution, 2013).

Table 2. Disadvantages of hydroponic culture

Disadvantage	Relevant facts
Cost	Requirement of high initial setup and constant supervision
	High investment in irrigation/automatic fertilizer systems and lights/indoor environment
Susceptibility to power outage	Necessity of manual watering
Risk of disease infection	Easy introduction of water-based microbes in closed system
No soil buffer	Possible crop failure
Relatively new cultivation technique	Requirement of technical knowledge with high-level management skills
Disease infection	In a "closed systems" or when excess drain water flows along the roots

1.3. Use as Research Tool

1.3.1. Bases of Hydroponics

Hydroponics have the potential to access all plant tissues and easy manipulation of the nutrient profile of the growth medium when compared with soil cultivation (Hershey, 1994; Jones, 1982). The use of hydroponic culture is unaffected by weather conditions. Hydroponics are advantageous for physiological, biochemical and molecular studies and allow easy observation and measurement of root growth and morphology throughout its life cycle (Arteca and Arteca, 2000; Grewal, 2011; Parks et al., 2009). Considering these merits, the soilless cultivation of plants serve exclusively as a tool for plant nutrition studies (Savvas, 2003). This method is free of confounding or uncontrollable variations in soil nutrient supply (Nhut et al. 2004). Accordingly, hydroponics is a useful tool to assess the feasibility of liquid fertilizers. Today, hydroponics is considered as a promising technique for plant physiology and pathology experiments as well as for commercial production, although appropriate maintenance is required to manage hydroponic systems (Malvick and Percich, 1993; Modularhydro, 2011; Nhut et al., 2004; Resh, 1993).

1.3.2. Practical Use

Hydroponics have been applied to plant nutrition studies. A hydroponic culture system was designed to study the effects of silicon on the growth of wild rice (*Zizania palustris* var. *interior* L.) and the resistance of wild rice to fungal brown spot disease. Mineral nutrient requirements of wild rice grown on flooded soils were not fully understood (Malvick and Percich, 1993). Plant and mycorrhizal responses to arsenic (As) have been investigated based on hydroponic experiments under a controlled environment (Fitz and Wenzel, 2002). Through this culture system, As-induced toxicity, As–P interactions, nutritional implications of excess As supply, uptake kinetics, bioavailability of different As species, and genetics of As-tolerance could be examined. Potassium (K), as the most abundant (3-5% of the total dry weight) cation in plants, is essential for plant growth. Although molecular details of the process involved in K^+ radial movement from the root surface to the xylem are not fully understood (Marschner, 1995; Tester and Leigh, 2001), it was suggested that root hairs play an important role in K^+ uptake using hydroponic culture system for growing Arabidopsis (Ahn et al., 2004). The effect of high zinc (Zn) concentration in the nutrient solution on the growth, photosynthetic characteristics and nutrient composition of sugar beet (*Beta vulgaris* L.)

was investigated using hydroponics (Sagardoy et al., 2009). Moreover, tolerance and accumulation of hazardous cadmium (Cd) have been investigated in rice under hydroponics, and manganese (Mn) could effectively control Cd accumulation and ameliorate Cd toxicity (Huang et al., 2017).

Hydroponic greenhouses have been designed to optimize the internal climate and the delivery of water and nutrients to growing plants, which can better control efficiency in water and nutrient use than field production systems (Bradley and Marulanda, 2000; Sheikh, 2006). The use of hydroponics provided precise control of internal climate and monitoring of water and nutrient parameters (Grewal et al., 2011). The efficiency of this system was dependent on the reuse of drainage water containing nutrients contributing to the pollution of local waterways (Thompson et al., 2007). Furthermore, hydroponics have been used to study the effect of nitrogen levels on the growth of spearmint (*Mentha spicata* L.) and its biochemical characteristics including chlorophyll content, and antioxidant and ascorbate peroxidase activities (Chrysargyris et al., 2017). Under a nitrogen-free hydroponics condition, the effect of plant-growth-promoting Rhizobacterial inoculation on growth and N_2 fixation of tissue-cultured banana plantlets was evaluated to apply biological N_2-fixation technology to a high N-fertilizer requiring plant (Baset et al., 2010). Recently, hydroponics has been used to assess the feasibility of biodegraded fishery wastewater as a liquid fertilizer, and to exploit resources from organic wastewater (Dao and Kim, 2011).

2. APPLICATION OF HYDROPONIC CULTURE

2.1. Case Study in Reuse of Fishery Waste

In 2014, global fishery production reached 167.2 million tons by capture (93.4 million tons) and by aquaculture (73.8 million tons) (Nhut et al., 2006). Human consumption was 146.3 million tons, corresponding to 20.1 kg per capita food. The increase in fishery production is the result of an increase in population. This situation yields a great deal of fishery waste that has not been efficiently reused. Therefore, the ideal reuse of fishery waste remains a matter of concern and interest.

2.1.1. Fish Waste

So far, solid fish waste has been mostly recycled to produce fishmeal or treated together with municipal waste, while wastewater generated from fish processing has been disposed of via municipal sewage system or directly into a waterbody. For wastewater disposal, the receiving waterbody must adequately degrade wastewater components without damaging the relevant ecosystem (FAO, 2005). Liquid fertilizers have recently been proposed as methods for fish waste/wastewater reuse (Dao and Kim, 2011; Figueroa et al., 2015; Gwon and Kim, 2012; Jung and Kim, 2016). Fish protein hydrolysates (FPH) and hydrolyzed fish materials are good sources of fertilizers for plants, and their effectiveness is related to their amino acid composition and rapid uptake through roots and leaves (Kristinsson, 2007). In addition, FPH were found to greatly stimulate the production of certain valuable bioactive compounds beneficial for plant growth. For instance, the addition of mackerel FPH in growth media resulted in an increase in the production of rosemaric acid and phenolics in oregano (Andarwlan and Shetty, 1999). In anise root cultures, the mackerel FPH increased the production of phenolic compounds that could help regulate the production of major phytochemicals in plants (Andarwlan and Shetty, 1999). Increased phenolic compounds in cranberry pomace was also reported when FPH was added to the growth media (Vattem and Shetty, 2002). Furthermore, amino acids, such as tryptophan, methionine, and cysteine, have been shown to have antioxidant capacity, and have beneficial effects on the growth of sunflower (*Helianthus annuus* L.) (Al-Qubaie, 2012). Therefore, the functions of FPH in the plant culture could go beyond providing basic nutrition, and could lead to the development of liquid fertilizer from fish waste/wastewater.

Table 3 represents the reported results of hydroponic culture using various fish waste sources, compared with the results of hydroponic culture using commercial organic fertilizers. Although the growth rate of each plant is somewhat different, its eventual growth on various fish waste sources is comparable to that on commercial fertilizers, with the exception of red bean cultivated on the supernatant of anchovy wastewater. Red bean is presumed to be sensitive to the salt content in anchovy supernatant, since salt is thought to affect nitrogen and phosphorus metabolism in hydroponic culture (Parida and Das, 2004). The discrepancies in plant growth rate shown in hydroponics may be caused by components of nutrition solution, culture conditions, health state of plant seed, etc. (Hussain et al., 2014; Modularhydro, 2011; Sardare and Admane, 2013). The culture supernatant obtained from the biodegraded fish waste or wastewater is a rich source of amino acids. Table 4 represents amino acids compositions obtained from the different fish waste sources, compared with those obtained from commercial fertilizers. Although the level of each amino acid and the total amount were slightly different in various fish waste sources, the components were comparable to commercial fertilizers. Therefore, the amino acid compositions were somewhat related to the effectiveness of the fertilizer, which could be conveniently investigated through hydroponic culture.

As shown in Table 3, the culture broths (including supernatant and microbes) of mackerel wastewater, as biofertilizer showed good growth in both barley and red bean, which was comparable to commercial fertilizers. Recently, development of biofertilizers from fish waste/wastewater has aroused interest. The complete reuse of fish waste/wastewater as biofertilizer (composed of microbes) following biodegradation is economically worthwhile. Microbial activity increases during biodegradation, and the remaining cells after biodegradation must be disposed of unless they are used together with liquid culture supernatant. To dispose of the remaining microbes (chemical composition is represented as $C_5H_7O_2N$ in general), this organic compound should be oxidized. For complete oxidation, two types of oxygen are needed: the carbonaceous oxygen demand to oxidize organic carbon to carbon dioxide, and the nitrogenous oxygen demand to convert ammonia to nitrate (Masters and Ela, 2008). Therefore, the entire biodegraded fish waste/wastewater used as a biofertilizer provides a benefit by reduction in expenditure for cell treatment. Meanwhile, microbes included in the biofertilizer should have no detrimental effect on plants; otherwise it cannot be used together with supernatant culture broth. As shown in Table 3, in a closed hydroponic culture system, biodegraded mackerel wastewater can be used as a biofertilizer on barley and red bean (Jung and Kim, 2016). The results of the hydroponic culture indicated that mackerel wastewater-degrading microbes (mainly *Bacillus* species) may have useful functions, such as nitrogen fixation, potassium and phosphorus solubilization, and synthesis of growth-promoting substances (Mohammadi and Sohrabi, 2012). In addition, such plant-growth-promoting *Bacillus* species can greatly reduce susceptibility to diseases and improve tolerance to environmental stress (Khan et al., 2011; Nautiyal et al., 2013; Siddikee et al., 2011).

2.1.2. Seaweed Waste

Seaweeds are classified into three major groups: red (*Rhodophyta*), brown (*Phaeophyta*), and green (*Chlorophyta*). Their carbohydrate content of red, brown and green seaweeds range between 40-75%, 36-60% and 41-53% (based on dry weight), respectively (Suo et al., 1986). Content based on dry weight of proteins and lipids are 2-39% and 0-2% (in red seaweed), 6-20% and 1-3% (in brown seaweed), and 17-23% and 0-1% (in green seaweed), respectively. For this reason, seaweeds have long been used as soil fertilizers, and spraying seaweed extracts have been reported to have positive effects on various crop plants; better seed germination, higher yields, stronger resistance to pests and diseases and longer shelf life of fruits (Blunden, 1991; Hankins and Hockey, 1990; Jolivet et al., 1991; Stephenson, 1966).

Oligomers degraded from sodium alginate (polysaccharide of brown seaweed) by gamma-ray irradiation have been shown to stimulate diverse biological and physiological activities (Aftab et al., 2011; Hien et al., 2000; Khan et al., 2011; Naeem et al., 2012a; Naeem et al., 2012b; Sarfaraz et al., 2011). As plant growth promoting substances, oligomers are thought to be involved in seed germination (Hien et al., 2000), shoot elongation (Hien et al., 2000; Natsume et al., 1994), root growth (Iwasaki and Matsubara 2000), flower production, antimicrobial activity, amelioration of heavy

Table 3. Results of hydroponic culture on various fish waste sources, compared with those on commercial fertilizers

Sources of fertilizer	Growth of plants after each culture day (cm)																References
	Barley								Red bean								
	0	2	3	7	8	9	10	14	0	2	3	7	8	9	10	14	
Mackerel (supernatant)	0	-[a]	0.3	-	-	7.0	-	11.1	0	-	1.4	-	-	5.5	-	9.8	[H1]
Mackerel (supernatant with cells)	0	-	0.1	-	-	4.8	-	11.2	0	-	1.7	-	-	6.7	-	11.6	[H1]
Anchovy (supernatant)	0	0.3	-	-	7.0	-	-	11.6	0	0	-	-	2	-	-	5.5	[H2]
Fish mixture (supernatant)	0	-	3.5	4.0	-	-	4.5	-	0	-	1.2	7.0	-	-	8.3	-	[H3]
Commercial product 1	0	-	3.2	3.8	-	-	4.5	-	0	-	1.2	6.3	-	-	7.8	-	[H3]
Commercial product 2	0	-	3.5	4.0	-	-	4.5	-	0	-	1.2	7.0	-	-	-	8.3	[H3]
Commercial product 3	0	0	-	-	5.2	-	-	11.2	0	0.5	-	-	3.75	-	-	7.0	[H2]

[a] -: not measured.

Table 4. Amino acid compositions (g 100 g^{-1}) in various fertilizer sources

Amino acids	Fish waste sources			Commercial fertilizers[d]
	Waste[a]	Wastewater[b]	Anchovy wastewater[c]	
Aspartic acid	0.80	0.34	n.d.[e]	0.74
Threonine	0.23	0.78	n.d.	0.22
Serine	0.22	0.09	n.d.	0.22
Glutamic acid	0.86	0.35	0.35	1.93
Proline	0.50	0.11	n.d.	0.35
Glycine	0.92	0.32	0.31	0.78
Alanine	0.70	1.06	0.86	0.99
Valine	0.36	0.35	0.61	0.32
Isoleucine	0.23	0.51	0.53	0.21
Leucine	0.37	0.76	1.22	0.34
Tyrosine	0.12	0.24	1.39	0.11
Phenylalanine	0.22	0.52	0.49	0.20
Histidine	0.20	0.13	0.12	0.27
Lysine	0.73	0.18	0.64	1.04
Asparagine	n.d.	0.02	n.d.	n.d.
Glutamine	n.d.	0.11	0.30	n.d.
Arginine	0.31	0.38	n.d.	0.29
Cystine	0.06	0.24	0.43	0.02
Methionine	0.06	0.21	0.09	0.01
Tryptophan	0.02	0.60	0.70	0.01
Total	6.91	7.30	8.04	8.05

[a] 96-h culture supernatant of fish waste (Dao and Kim, 2011).
[b] 52-h culture supernatant of fishmeal wastewater (Gwon and Kim, 2012).
[c] 70-h culture supernatant of anchovy fishmeal wastewater (Figueroa et al., 2015).
[d] Average values of two commercial fertilizers (Figueroa et al., 2015).
[e] n.d.: not detected.

metal stress, synthesis of phytoalexins, etc. (Darvill et al., 1992; Hu et al., 2004; Kume et al., 2002; Luan et al., 2003). Laminarin, another polysaccharide of brown seaweed, and its depolymerized oligosaccharides are well known to be involved in plant protection by the induction of defense response and resistance against several pathogens (Aziz et al., 2003; Vera et al., 2011). Thus, these compounds are main components of several commercial seaweed liquid fertilizers. In addition, κ-carrageenans, a family of partly sulphated linear galactans found in the cell walls of red seaweed, have been reported to elicit β-1,3-glucanase activity in *Rubus fruticosus* cell suspension cultures, and oligo-κ-carrageenans were more efficient than the native polysaccharide in this study (Patier et al., 1995). Furthermore, small molecules degraded from carrageenans by irradiation could be an effective natural fertilizer, since they could promote growth in rice and increase resistance to certain pests (Keller, 2015). The biodegraded culture broth of seaweed also showed antioxidant, antibacterial, and antifungal activities, and the entire culture broth including microbes was potentially useful as a biofertilizer when applied to lab-scale hydroponics (Gupta and Abu-Ghannam, 2011; Zodape, 2001).

Nowadays, seaweed extracts have been used as liquid fertilizers. They are biodegradable, non-toxic, eco-friendly and non-hazardous to humans and animals, and therefore have advantages over chemical fertilizers (Dhargalkar and Pereira, 2005). It has been reported that some important major and minor nutrients were detected in seaweed extracts, which could enhance the growth of plants and secondary metabolite compositions (Elansary et al., 2016). *Vigna mungo* seeds soaked with a low concentration of seaweed extracts showed improved germination rate, although germination was inhibited at a high concentration of the extracts (Ganapathy Selvam et al., 2013). Therefore, seaweed extracts can increase yield and resistance of many crops by increasing germination rate. Moreover, it has been reported that application of seaweed extracts increased chlorophyll content of *Cyamopsis tetrogonolaba* (L.) Taub (Thirumaran et al., 2009).

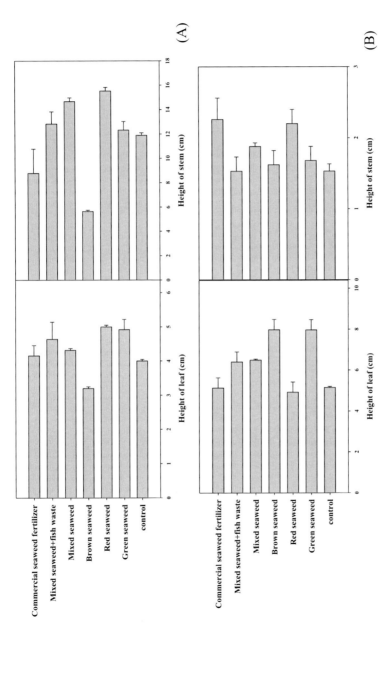

Figure 1. Results of 15-d hydroponic cultures of kidney bean (A) and barley (B) using 4-d culture supernatants of various seaweed wastes (mixture of seaweeds and fish wastes, mixture of seaweeds, brown seaweed, red seaweed and green seaweed wastes) and a commercial seaweed fertilizer. Data were extracted from the study of Kim et al. (2014).

Hydroponic cultures of kidney bean and barley were conducted using 4-d culture supernatants of various seaweed wastes, compared with commercial seaweed fertilizer (Figure 1). With the exception of the culture supernatant of brown seaweed waste, most culture supernatants of various seaweed wastes showed better leaf and stem growth of both plants when compared with control. This was most clear in the hydroponic culture of kidney bean, and the fertilizing ability of most seaweed culture supernatants was superior to commercial seaweed fertilizer. Fertilizing ability is related to the amino composition of fertilizers. Amino acids are essential and the precursors of several metabolites that regulate plant growth (Amir et al., 2002). Therefore, discrepancies in fertilizing ability may be related to total amino acids and/or concentration of a specific amino acid (Table 5).

Table 5. Amino acid composition in the biodegraded culture broths of various seaweed wastes[a]

Amino acid	Seaweed waste sources (g 100g^{-1})		
	Green seaweed	Red seaweed	Brown seaweed
Aspartic acid	0.05	0.06	0.06
Threonine	n.d.[b]	n.d.	n.d.
Serine	n.d.	n.d.	n.d.
Glutamic acid	n.d.	n.d.	0.24
Proline	n.d.	n.d.	0.04
Glycine	0.03	0.07	0.06
Alanine	n.d.	n.d.	0.23
Valine	0.01	0.06	0.05
Isoleucine	n.d.	n.d.	0.03
Leucine	0.12	n.d.	0.07
Tyrosine	1.49	n.d.	0.03
Phenylalanine	2.6	n.d.	0.39
Histidine	0.22	n.d.[a]	n.d.
Lysine	3.22	0.70	0.06
Asparagine	n.d.	n.d.	n.d.
Glutamine	0.41	0.64	n.d.
Arginine	n.d.	n.d.	n.d.
Cysteine	n.d.	0.48	n.d.
Methionine	0.78	0.46	n.d.
Tryptophan	n.d.	n.d.	n.d.
Total	8.93	2.47	1.26

[a] Data were extracted from the study of Kim et al. (2014).
[b] n.d.: not detected.

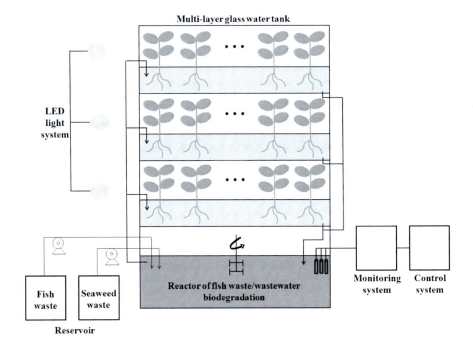

Figure 2. A building-type hydroponic culture system using biofertilizer manufactured from fish waste/wastewater.

2.2. Proposed Model

Hydroponics are a potential method of growing plants without soil, by exposing the roots to mineral nutrient solution. A good source of nutrients used in hydroponics can be obtained from fish waste/wastewater. Using biofertilizer manufactured from biodegraded fish waste/wastewater, an advanced building-type hydroponic culture system is proposed to cultivate fruits and vegetables for urban consumption (Figure 2).

2.2.1. Characteristics of a Novel Hydroponics Model

Hydroponic culture makes the most efficient use of available space, since plants can be stacked in many configurations (Sonneveld, 2000). Multi-layer water tanks are designed in the proposed model. At the base tank, nutrition including bioactive compounds beneficial for plant growth is produced by biodegradation using microbes. The substrates (fish and seaweed waste) are pumped from a reservoir, the nutrition solution is pumped into each water tank, and the nutrition solution returns to the base tank. The mixed-type substrates are designed for biodegradation in this model, since it is difficult to segregated collection of diverse fishery waste. The amount of nutrient solution pumping into each water tank should be sufficient to prevent the plants from drying out. In the base tank, reaction monitoring and control systems are installed to maintain stable biodegradation, and substrate pumping rate is adjusted according to the quality of nutrition and natural evaporation. LED lights are provided for plants growing in each layer (Xu et al., 2016).

This newly proposed hydroponic culture system can provide useful data to construct a real building-type hydroponics system. Through this research tool, major parameters determining the success in operation of this hydroponics system can be obtained. If this hydroponics system is feasible, full reuse of mixed-type fishery waste/water is possible in an eco-friendly way, without producing other waste or wastewater during the treatment processes. Using this hydroponics system, production of fruits and vegetables as well as disposal of fishery waste/wastewater is practical. To operate this hydroponics system, development of related technologies is essential, increasing production compared with soil cultivation (Ghehsareh et al., 2011; Olivia's growing solution, 2013; Parks et al., 2009). This hydroponics system can utilize urban buildings. The waste fish and seaweeds, which are consumed in urban areas, are disposed of and reused as biofertilizer by this hydroponics system in a continuous fashion. In the same place, fruits and vegetables can be cultivated and produced, reducing distribution costs greatly.

2.2.2. Possible Problems of this Model

Since this hydroponic culture system is new, possible problems may occur during operation, and should be considered for practical use. Firstly, any phytotoxic compounds may be produced at the base water tank where biodegradation occurs, damaging plant growth. Although seafood processing wastewater was reported to be nontoxic (Afonso and Borquez,

2002), intermediate metabolites produced from the mixed-type fish or seaweed waste/wastewater during the biodegradation may be phytotoxic. Secondly, it may be difficult to maintain nutrient quality at the base water tank. Although high and stable production of bioactive compounds from the mixed-type fish or seaweed waste/wastewater is desirable in this system, inferior results can be caused by poor biodegradation. This can be due to an imbalance in C/N ratio of substrates, enzyme inhibition by intermediate metabolites, and variance in dilution effect between evaporation and circulating water input (Wu et al., 2017). Thirdly, pathogens may be detected at the base water tank. Under an open hydroponic culture system, there is possible infection of water-borne pathogens, which can spread along with the circulating water. The pathogens compete with useful microbes dominating in the base water tank. Accordingly, the nutrient quality would deteriorate through poor biodegradation as pathogens proliferate. Furthermore, these pathogens can cause severe damage to the growing plants (Olympios, 1999). Finally, there may be concerns on the application of small-scale data to building-type hydroponic culture systems. Difficulty in scaling-up may occur unless major parameters of small-scale hydroponics are applicable.

CONCLUSION

Nowadays, eco-friendly zero-emission management of fishery waste arouses interest to turn fishery waste into valuable resources including liquid fertilizers. When fishery waste is biodegraded to expand reuse, the resultant culture broth containing some bioactive metabolites and microbes can be used as a biofertilizer. To assess the feasibility of the culture broth as biofertilizer, hydroponics can be selected as a useful tool. The success of lab-scale hydroponics using fishery waste has led to a design for an advanced building-type hydroponic culture system to cultivate fruits and vegetables for urban consumption. For practical use of this hydroponics model proposed in this chapter, more efforts are required with the development of hydroponics technology.

ACKNOWLEDGMENT

This research was supported by a grant (Receipt number: 2017019021, SGER 2017) from the National Research Foundation of Korea.

REFERENCES

Afonso, M.D. and Borquez, R. (2002). Review of the treatment of seafood processing wastewaters and recovery of proteins therein by membrane separation processes - prospects of the ultrafiltration of wastewaters from the fish meal industry. *Desalination*, *142*, 29-45.

Aftab, T., Khan, M.M.A., Idrees, M., Naeem, M., Moinuddin, Hashmi, N. and Varshney, L. (2011). Enhancing the growth, photosynthetic capacity and artemisinin content in *Artemisia annua* L. by irradiated sodium alginate. *Radiat. Phys. Chem.*, *80*, 833-836.

Ahluwalia, S.S. and Goyal, D. (2007). Microbial and plant derived biomass for removal of heavy metals from wastewater, *Bioresour. Technol.*, *98*, 2243-2257.

Ahn, S.J., Shin, R. and Schachtman, D.P. (2004). Expression of *KT/KUP* genes in Arabidopsis and the role of root hairs in K^+ uptake. *Plant Physiol.*, *134*, 1135-1145.

Al- Qubaie, A.I. (2012). Response of sunflowers cultivar Giza- 102 (*Helianthus annuus, L*) plants to spraying some antioxidants. *Nat. Sci.*, *10*, 1-6.

Allègre, M., Héloir, M. C., Trouvelot, S., Daire, X., Pugin, A., Wendehenne, D. and Adrian, M. (2009). Are grapevine stomata involved in the elicitor-induced protection against downy mildew? *Mol. Plant Microbe Interact*, *22*, 977-986.

AlShrouf, A. (2017). Hydroponics, aeroponic and aquaponic as compared with conventional farming. *ASRJETS, 27(1)*, 247-255.

Amir, R., Hacham, Y. and Galili, G. (2002). Cystathionine γ-synthase and threonine synthase operate in concert to regulate carbon flow towards methionine in plants. *Trends Plant Sci.*, *7*, 153-156.

Andarwlan, N. and Shetty, K. (1999). Influence of acetyl salicylic acid in combination with fish protein hydrolysates on hyperhydricity reduction and phenolic synthesis in oregano (*Origanum vulgare*) tissue cultures. *J. Food Biochem., 23*, 619-635.

Angelakis, A., Do Monte, M.M., Bontoux, L. and Asano, T. (1999). The status of wastewater reuse practice in the Mediterranean basin: need for guidelines. *Water Res., 33*, 2201-2217.

Arteca, R.N. and Arteca, J.M. (2000). A novel method for growing *Arabidopsis thaliana* plants hydroponically. *Physiol. Plant, 108*, 188-193.

Association for Vertical Farming (AVF) (2015). URL: https://vertical-farming.net.

Aziz, A., Poinssot, B., Daire, X., Adrian, M., Bézier, A., Lambert, B., Joubert, J.M. and Pugin, A. (2003). Laminarin elicits defense responses in grapevine and induces protection against *Botrytis cinerea* and *Plasmopara viticola*. *Mol. Plant Microbe Interact., 16*, 1118-1128.

Baset, M., Shamsuddin, Z.H., Wahab, Z. and Marziah, M. (2010). Effect of plant growth promoting Rhizobacterial (PGPR) inoculation on growth and nitrogen incorporation of tissue-cultured 'Musa' plantlets under nitrogen-free hydroponics condition. *Aust. J. Crop Sci., 4(2)*, 85-90.

Beibel, J.P. (1960). *Hydroponics -The Science of Growing Crops Without Soil*. Florida: Florida Department of Agric. Bull.

Bellona Foundation (2009). The Sahara forest project. URL: http://www.saharaforestproject.com/#.

Besthorn, F.H. (2013). Vertical farming: Social work and sustainable urban agriculture in an age of global food crises. *Aust. Soc. Work, 66*, 187-203.

Black, K. (2016). The disadvantages of hydroponics. URL: http://www.gardenguides.com/75433-disadvantages-hydroponics.html.

Blunden, G. (1991). Agricultural uses of seaweeds and seaweed extracts. In: Guiry, M.D. and Blunden, G. (eds.), *Seaweed Ressources in Europe: Uses and Potential*. Chichester, UK: John Wiley & Sons Ltd, pp. 65-81.

Bonhoff, A. and Grisebach, H. (1988). Elicitor-induced accumulation of glyceollin and callose in soybean roots and localized resistance against *Phytophthora megasperma* f. sp. *glycinea*. *Plant Sci.*, *54*, 203-209.

Bowring, S. (2012). Mission 2015: Hydroponic culture. URL: http://web.mit.edu/12.000/www/m2015/2015/hydro_agriculture.html.

Bradley, P. and Marulanda, C. (2000). Simplified hydroponics to reduce global hunger. *Acta. Hort.*, *554*, 289-295.

Butler, J.D., and Oebker, N.F. (2006). *Hydroponics as a Hobby— Growing Plants Without Soil*. Illinois: University of Illinois, College of Agriculture, Extension Service in Agriculture and Home Economics.

Cavaliero, G. (2011). What is aquaponics? URL: https://www.theaquaponicsource.com/what-is-aquaponics.

Charles, H. and Godfray, J. (2011). Food and biodiversity. *Science, 333*, 1231-1232.

Chekli, L., Kim, J.E., El Saliby, I., Kim, Y., Phuntsho, S., Li, S., Ghaffour, N., Leiknes, T. and Shon, H.K. (2017). Fertilizer drawn forward osmosis process for sustainable water reuse to grow hydroponic lettuce using commercial nutrient solution. *Sep. Purif. Technol.*, In Press.

Chrysargyris, A., Nikolaidou, E., Stamatakis, A. and Tzortzakis, N. (2017). Vegetative, physiological, nutritional and antioxidant behavior of spearmint (*Mentha spicata* L.) in response to different nitrogen supply in hydroponics. *J. Appl. Res. Med. Aromat. Plants*, In Press.

Côté, F., Ham, K.S., Hahn, M.G. and Bergmann, C.W. (1998). Oligosaccharide elicitors in host–pathogen interactions: generation, perception and signal transduction. In: Biswas, B.B. and Das, H.K. (eds.), *Subcellular Biochemistry, Plant–Microbe Interactions*. New York, USA: Plenum Press, pp. 385-432.

Dao, V. T. and Kim, J. K. (2011). Scaled-up bioconversion of fish waste to liquid fertilizer using a 5 L ribbon-type reactor. *J. Environ. Manage.*, *92*, 2441-2446.

Darvill, A., Augur, C., Bergmann, C., Carlson, R.W., Cheong, J.J., Eberhard, S., Hahn, M.G., Lo, V.M., Marfa, V., Meyer, B., Mohnen, D., O'Neill, M.A., Spiro, M.D., van Halbeek, H., York, W.S. and Albersheim, P. (1992). Oligosaccharins--oligosaccharides that regulate growth, development and defence responses in plants. *Glycobiology*, 2, 181-198.

de Anda, J. and Shear, H. (2017). Potential of vertical hydroponic agriculture in Mexico. *Sustainability*, 9, 1-17.

De Ruiter, G.A. and Rudolph, B. (1997). Carrageenan biotechnology. *Trends Food Sci. Tech.*, 8, 389-395.

Despommier, D. (2009). The rise of vertical farms. *Sci. Am.*, 301, 80-87.

Despommier, D. (2010). *The Vertical Farm: Feeding the World in the 21st Century*, New York, Picador.

Dhargalkar, V.K. and Pereira, N. (2005). Seaweed: promising plant of the millennium. *Science*, 71, 60-66.

Diez, R. (2015). Hydroponics: advantages and disadvantages. URL: http://dyna-gro-blog.com/hydroponics-advantages-and-disadvantages.

Dregne, H.E. (2002). Land degradation in the drylands. *Arid Land Res. Manag.*, 16, 99-132.

Elansary, H.O., Norrie, J., Ali, H.M., Salem, M.Z.M., Mahmoud, E.A. and Yessoufou, K. (2016). Enhancement of *Calibrachoa* growth, secondary metabolites and bioactivity using seaweed extracts. *BMC Complement. Altern. Med.*, 16, 1-11.

Ellis, N.K., Jensen, M., Larsen, J., and Oebker, N. (1974). *Nutriculture Systems—Growing Plants Without Soil*. Indiana: Purdue University, Station Bulletin (44).

Facinor, J. (2016). Knowing more about aquaponics vs hydroponics. URL: http://facinor.site/knowing-more-about-aquaponics-vs-hydroponics.

FAO (2005). Waste management of fish and fish products. URL: http://www.fao.org/fishery/topic/12326/en.

Färe, R., Grosskopf, S. and Weber, W.L. (2006). Shadow prices and pollution costs in U.S. agriculture. *Ecol. Econ.*, 56, 89-103.

Ferro, G., Fiorentino, A., Alferez, M.C., Polo-López, M.I., Rizzo, L. and Fernández-Ibánez P. (2015). Urban wastewater disinfection for agricultural reuse: effect of solar driven AOPs in the inactivation of a multidrug resistant *E. coli* strain. *Appl. Catal., B, 178*, 65-73.

Figueroa, J.G.S., Jung, H.Y., Jeong, G.T. and Kim, J.K. (2015). The high reutilization value potential of high-salinity anchovy fishmeal wastewater through microbial degradation. *World J. Microbiol. Biotechnol., 31,* 1575-1586.

Fitz, W.J. and Wenzel, W.W. (2002). Arsenic transformations in the soil-rhizosphere-plant system: fundamentals and potential application to phytoremediation. *J. Biotechnol., 99(3),* 259-278.

Fornes, F., Sanchez-Perales, M. and Guadiola, J.L. (2002). Effect of a seaweed extract on the productivity of 'de Nules' Clementine mandarin and navelina orange. *Bot. Mar., 45,* 486-489.

Ganapathy Selvam, G., Balamurugan, M., Thinakaran, T. and Sivakumar, K. (2013). Developmental changes in the germination, growth and chlorophyllase activity of *Vigna mungo* L. using seaweed extract of *Ulva reticulata* Forsskal. *Intn. Res. J. Pharm., 4,* 252-254.

Ghehsareh, A.M., Khosravan, S. and Shahabi, A.A. (2011). The effect of different nutrient solutions on some growth indices of greenhouse cucumber in soilless culture. *J. Plant Breed. Crop Sci., 3(12),* 321-326.

Grewal, H.S., Maheshwari, B. and Parks, S.E. (2011). Water and nutrient use efficiency of a low cost hydroponic greenhouse for a cucumber crop: An Australian case study. *Agric. Water Manage., 98,* 841-846.

Gruda, N. (2009). Do soilless culture systems have an influence on product quality of vegetables? *J. Appl. Bot. Food Q., 82,* 141-147.

Gupta, S. and Abu-Ghannam, N. (2011). Recent developments in the application of seaweeds or seaweed extracts as a means for enhancing the safety and quality attributes of foods. *Innov. Food Ssci. Emerg. Technol.,* 12, 600-609.

Gwon, B.G. and Kim, J.K. (2012). Feasibility study on production of liquid fertilizer in a $1m^3$ reactor using fishmeal wastewater for commercialization. *Environ. Eng. Res., 17,* 3-8.

Hahn, M.G. (1996). Microbial elicitors and their receptors in plants. *Annu. Rev. Phytopathol., 34, 387-412.*

Halais, F. (2014). Can urban agriculture work on a commercial scale? URL: http://citiscope.org/story/2014/can-urban-agriculture-work-commercial-scale.

Hankins, S.D. and Hockey, H.P. (1990). The effect of a liquid seaweed extract from Ascophyllum nodusum (Fucales, Phaeophyta) on the two-spotted red spider mite Tetranychus urticae. *Hydrobiologia, 204/205,* 555-559.

Hershey, D.R. (1994). Solution culture hydroponics: history and inexpensive equipment. *Am. Biol. Teacher., 56,* 111-118.

Hien, N.Q., Nagasawa, N., Tham, L.X., Yoshii, F., Dang, H.V., Mitomo, H., Makuuchi, K. and Kume, T. (2000). Growth promotion of plants with depolymerised alginates by irradiation. *Radiat. Phys. Chem., 59,* 97-101.

Home Hydro Systems (2015). Aeroponic system. URL: http://www.homehydrosystems.com/hydroponic-systems/aeroponics_systems.html.

Hu, X., Jiang, X., Hwang, H., Liu, S. and Guan, H. (2004). Promotive effects of alginate-derived oligosaccharide on maize seed germination. *J. Appl. Phycol., 16,* 73-76.

Huang, Q., An, H., Yang, Y.J., Liang, Y. and Shao, G.S. (2017). Effects of Mn-Cd antagonistic interaction on Cd accumulation and major agronomic traits in rice genotypes by different Mn forms. *Plant Growth Regul.,* 1-15.

Hussain, A., Iqbal, K., Aziem, S., Mahato, P. and Negi, A.K. (2014). A review on the science of growing crops without soil (soilless culture) − A novel alternative for growing crops. *Intl. J. Agri. Crop. Sci., 7(11),* 833-842.

Ibrahim, M.H., Jaafar, H.Z., Karimi, E. and Ghasemzadeh, A. (2012). Primary, secondary metabolites, photosynthetic capacity and antioxidant activity of the Malaysian Herb Kacip Fatimah (*Labisia pumila* Benth) exposed to potassium fertilization under greenhouse conditions. *Int. J. Mol. Sci., 13,* 15321-15342.

Iwasaki, K. and Matsubara, Y. (2000). Purification of alginate oligosaccharides with root growth-promoting activity toward lettuce. *Biosci. Biotechnol. Biochem., 64,* 1067-1070.

Jolivet, E., Langlais-Jeannin, I. and Morot-Gaudry, J.F. (1991). Les extraits d'algues marines: propriétés phytoactives et intérêt agronomique. *L' Année Biologique, 30,* 109-126.

Jones, J.B., Jr (1982). Hydroponics: Its history and use in plant nutrition studies. *J. Plant Nutr., 5,* 1003-1030.

Jung, H.Y. and Kim, J.K. (2016). Eco-friendly waste management of mackerel wastewater and enhancement of its reutilization value. *Int. Biodeterior. Biodegrad., 111,* 1-13.

Keen, N.T., Yoshikawa, M. and Wang, M.C. (1983). Phytoalexin elicitor activity of carbohydrates from *Phytophthora megasperma* f.sp. *glycinea* and other sources. *Plant Physiol., 71,* 466-471.

Keller, R. (2015). Seaweed component found to be a rice fertilizer. URL: http://www.agprofessional.com/news/seaweed-component-found-be-rice-fertilizer.

Khan, N., Mishra, A., Chauhan, P.S. and Nautiyal, C.S. (2011). Induction of *Paenibacillus lentimorbus* biofilm by sodium alginate and $CaCl_2$ alleviates drought stress in chickpea. *Ann. App. Biol., 159,* 372-386.

Khan, Z.A., Khan, M.M.A., Aftab, T., Idrees, M. and Naeem, M. (2011). Influence of alginate oligosaccharides on growth, yield and alkaloid production of opium poppy (*Papaver somniferum* L.). *Front. Agric. China, 5,* 122-127.

Killebrew, K. and Wolff, H. (2010). Environmental impacts of agricultural technologies, Evans School Policy Analysis and Research, EPAR Brief. URL: http://econ.washington.edu/sites/econ/files/old-site-uploads/2014/06/2010-Environmental-Impacts-of-Ag-Technologies.pdf.

Kim, J.K., Kim, E.J. and Kang, K.H. (2014). Achievement of zero emissions by the bioconversion of fishery waste into fertilizer. In: Lopez-Valdez, F. and Fernandes-Luqueno, F. (eds.), *Fertilizers: components, uses in agriculture and environmental impacts.* New York, NY, USA: Nova Science Publishers, Inc., pp. 69-94.

Kipp, J. (2009). Optimal climate regions in Mexico for greenhouse crop production; Wageningen UR greenhouse horticulture. URL: http://mexico.nlambassade.org/binaries/content/assets/postenweb/m/m

exico/nederlandse-ambassade-in-mexico-stad/import/producten_en_diensten/landbouw/optimal-climate-regions-in-mexico-for-greenhouse-crop-production.

Klarzynski, O., Plesse, B., Joubert, J. M., Yvin, J. C., Kopp, M., Kloareg, B. and Fritig, B. (2000). Linear beta-1,3 glucans are elicitors of defense responses in tobacco. *Plant Physiol.*, *124*, 1027-1037.

Kobayashi, A., Tai, A., Kanzaki, H. and Kawazu, K. (1993). Elicitor-active oligosaccharides from algal laminaran stimulate the production of antifungal compounds in alfalfa. Z. Naturforsch., 48, 575-579.

Kristinsson, H.G. (2007). Functional and bioactive peptides from hydrolyzed aquatic food proteins. In: Barrow, C. and Shahidi, F. (eds.), *Marine nutraceuticals and functional foods*. New York, USA: CRC Press, pp. 229-246.

Kume, T., Nagasawa, N. and Yoshii, F. (2002). Utilization of carbohydrates by radiation processing. *Radiat. Phys. Chem.*, *63*, 625-627.

Luan, L.Q., Hien, N.Q., Nagasawa, N., Kume, T., Yoshii, F. and Nakanishi, T.M. (2003). Biological effect of radiation-degraded alginate on flower plants in tissue culture. *Biotechnol. Appl. Biochem.*, *38*, 283-288.

Malvick, D.K. and Percich J.A. (1993). Hydroponic culture of wild rice (*Zizania patu.srns* L.) and its application to studies of silicon nutrition and fungal brown spot disease. *Can. J. Plant Sci., 73(4)*, 969-975.

Marschner, H. (1995). *Mineral Nutrition of Higher Plants*. Massachusetts: Academic Press.

Masters, G.M. and Ela, W.P. (2008). *Introduction to Environmental Engineering and Science*, New Jersey: Pearson Prentice Hall.

Millennium Ecosystem Assessment (2005). Ecosystems and Human Well-Being. Desertification Synthesis. URL: http://millenniumassessment.org/documents/document.355.aspx.pdf.

Modularhydro (2011). Advantages of hydroponic soilless culture. URL:http://modularhydro.com/ArticleLibrary/AdvantagesOfHydroponicSoilessCulture.html.

Mohammadi, K. and Sohrabi, Y. (2012). Bacterial biofertilizers for sustainable crop production: A review. *ARPN J. Agri. Biol. Sci.*, *7*, 307-316.

Mudau, F.N., Soundy, P. and du Toit, E.S. (2007). Effects of nitrogen, phosphorus and potassium nutrition on total polyphenols content of bush tea (*Athrixia phylicoides* L.) leaves in a shaded nursery environment. *HortSci.*, *42*, 334-338.

Naeem, M., Idrees, M., Aftab, T., Khan, M.M.A., Moinuddin and Varshney, L. (2012a). Irradiated sodium alginate improves plant growth, physiological activities, and active constituents in *Mentha arvensis* L. *J. Appl. Pharm. Sci.*, *2*, 28-35.

Naeem, M., Idrees, M., Aftab, T., Khan, M.M.A., Moinuddin and Varshney, L. (2012b). Depolymerised carrageenan enhances physiological activities and menthol production in *Mentha arvensis* L. *Carbohydr. Polym.*, *87*, 1211-1218.

Naguib, A.M., El-Baz, F.K., Salama, Z.A., Hanna, H.A.E.B., Ali, H.F. and Gaafar, A.A. (2012). Enhancement of phenolics, flavonoids and glucosinolates of broccoli (*Brassica oleracea*, var. *Italica*) as antioxidants in response to organic and bio-organic fertilizers. *J. Saudi Soc. Agric. Sc.*, *11*, 135-142.

Natsume, M., Kamao, Y., Hirayan, M. and Adachi, J. (1994). Isolation and characterization of alginate derived oligosaccharides with root growth promoting activities. *Carbohydr. Res.*, *258*, 187-197.

Nautiyal, C.S., Srivastava, S., Chauhan, P.S., Seem, K., Mishra, A. and Sopory, S.K. (2013). Plant growth-promoting bacteria *Bacillus amyloliquefaciens* NBRISN13 modulates gene expression profile of leaf and rhizosphere community in rice during salt stress. *Plant Physiol. Biochem.*, *66*, 1-9.

Nhut, D.T., Dieu Huong, N.T. and Khiem, D.V. (2004). Direct microtuber formation and enhanced growth in the acclimatization of in vitro plantlets of taro (*Colocasia esculenta* spp.) using hydroponics. *Sci. Hortic.*, *101*, 207-212.

Nhut, D.T., Nguyen, N.H. and Thuy, D.T.T. (2006). A novel in vitro hydroponic culture system for potato (*Solanum tuberosum* L.) microtuber production. *Sci. Hortic.*, *110*, 230-234.

Olivia's growing solution (2013). Advantages & disadvantages of hydroponics! URL: http://www.oliviassolutions.com/blog/advantages-disadvantages-of-hydroponics.

Olympios, C.M. (1999). Overview of soilless culture: Advantages, constraints and perspectives for its use in Mediterranean countries. *Cahiers Options Méditerranéenn, 31,* 307-324.

Os, E.V., Gieling, T.H. and Ruijs, M.N.A. (2002). Equipment for hydroponic installations. In: Savvas, D. and Passam H.C. (eds.), *Hydroponic Production of Vegetables and Ornamentals*. Athens, Greece: Embryo Publications, PP. 103-141.

Parida, A.K. and Das, A.B. (2004). Effect of NaCl stress on nitrogen and phosphorus metabolism in a true mangrove *Bruguiera parviflora* grown under hydroponic culture. *J. Plant Physiol.*, *161*, 921-928.

Parks, S.E., Worrall, R.J., Low, C.T. and Jarvis, J.A. (2009). Initial efforts to improve the management of substrates in greenhouse vegetable production in Australia. *Acta Hort. 819,* 331-336.

Patier, P., Potin, P., Rochas, C., Kloareg, B., Yvin, J.C. and Liénart, Y. (1995). Free or silica-bound oligokappa-carrageenans elicit laminarase activity in Rubus cells and protoplasts. *Plant Sci.*, *110*, 27-35.

Patier, P., Yvin, J.C., Kloareg, B., Liénard, Y. and Rochas, C. (1993). Seaweed liquid fertilizer from Ascophyllum nodosum contains elicitors of plant D-glycanases. J. Appl. Phycol., 5, 343-349.

Podmirseg, D. (2014). Contribution of vertical farms to increase the overall energy efficiency of urban agglomerations. *J. Power Energy Eng.*, *2*, 82-85.

Potin P, Bouarab K, Kupper F and Kloareg B. (1999). Oligosaccharide recognition signals and defense reactions in marine plant–microbe interactions. *Curr. Opin. Microbiol.*, *2, 276-283.*

Pual, C. (2000). Heath and hydroponic. *Pract. Hydroponic Greenhouse, 53(4),* 28-30.

Putra, P.A. and Yuliando, H. (2015). Soilless culture system to support water use efficiency and product quality: A review. *Agric. Agric. Sci. Procedia, 3*, 283-288.

Qadir, M., Sharma, B.R., Bruggeman, A., Choukr-Allah, R. and Karajeh, F. (2007). Non-conventional water resources and opportunities for water augmentation to achieve food security in water scarce countries. *Agric. Water Manage., 87*, 2-22.

Resh, H.M. (1993). Water culture. In: Howard, M.R. (ed.), *Hydroponic Food Production*. Anaheim, California: Woodbridge Press Publishing Company, PP. 110-199.

Reynolds, J.F., Smith, D.M., Lambin, E.F., Turner, B.L., Mortimore, M., Batterbury, S.P., Downing, T.E., Dowlatabadi, H., Fernández, R.J., Herrick, J.E., Huber-Sannwald, E., Jiang, H, Leemans, R., Lynam, T., Maestre, F.T., Ayarza, M. and Walker, B. (2007). Global Desertification: Building a Science for Dryland Development. *Science, 316*, 847-851.

Sagardoy, R., Morales, F., Lo´pez-Milla´n, A.F., Abadı´a A. and Abadı´a J. (2009). Effects of zinc toxicity on sugar beet (*Beta vulgaris* L.) plants grown in hydroponics. *Plant Biol., 11*, 339-350.

Santos, J.D., Silva, A.L.L.D., Costa, J. D.L., Scheidt, G.N., Novak, A.C., Sydney, E.B. and Soccol, C.R. (2013). Development of a vinasse nutritive solutions for hydroponics. *J. Environ. Manage., 114*, 8-12.

Sardare, M.D. and Admane, S.V. (2013). A review on plant without soil – hydroponics. *IJRET, 2(3)*, 299-304.

Sarfaraz, A., Naeem, M., Nasir, S., Idrees, M., Atab, T., Hashmi, N., Khan, M.M.A., Moinuddin and Varshney, L. (2011). An evaluation of the effects of irradiated sodium alginate on the growth, physiological activity and essential oil production of fennel (*Foeniculum vulgare* Mill.). *J. Med. Plants Res., 5*, 15-21.

Savvas, D. (2003). Hydroponics: A modern technology supporting the application of integrated crop management in greenhouse. *JFAE, 1(1)*, 80-86.

Sheikh, B.A. (2006). Hydroponics: key to sustain agriculture in water stressed and urban environment. *Pak. J. Agric., Agril. Eng., Vet. Sci.*, *22*, 53-57.

Siddikee, M.A., Glick, B.R., Chauhan, P.S., Yima, W.J. and Sa, T. (2011). Enhancement of growth and salt tolerance of red pepper seedlings (*Capsicum annuum* L.) by regulating stress ethylene synthesis with halotolerant bacteria containing 1-aminocyclopropane-1-carboxylic acid deaminase activity. *Plant physiol. Biochem.*, *49*, 427-434.

Skyer, M. (2014). Vertical farming: It's coming to save the day, but will it? URL: http://www.craftsy.com/blog/2014/06/what-is-vertical-farming.

Song, W., Zhou, L., Yang, C., Cao, X., Zhang, L. and Liu, X. (2004). Tomato Fusarium wilt and its chemical control strategies in a hydroponic system. *Crop Prot., 23(3)*, 243-247.

Sonneveld, C. (2000). Effects of salinity on substrate grown vegetables and ornamentals in greenhouse horticulture. URL: http://edepot.wur.nl/121235.

Stephenson, W.M. (1966). The effect of hydrolysed seaweed on certain plant pests and diseases. *Proceedings of the International Seaweed Symposium, 5,* 405-415.

Suo, R.Y., Liu, G.Y., Shi, Z.J., Wang, F.J., Zhang, S. and Liu, T.J. (1986). Cultural test with submerged floating rope on giant kelp in South Huangcheng island. *Marine Fisheries Research*, *7*, 1-7.

Tester, M. and Leigh, R.A. (2001). Partitioning of nutrient transport processes in roots. *J. Exp. Bot.*, *52*, 445-457.

Thirumaran, G., Arumugam, M., Arumugam, R. and Anantharaman, P. (2009). Effect of seaweed liquid fertilizer on growth and pigment concentration of *Cyamopsis tetrogonolaba* L. Taub. *Am- Euras. J. Agron.*, *2*, 50-56.

Thompson, R.B., Martinez-Gaitan, C., Gallardo, M., Gimenez, C. and Fernandez, M.D. (2007). Identification of irrigation and N management practices that contribute to nitrate leaching loss from an intensive vegetable production system by use of a comprehensive survey. *Agric. Water Manage.*, *89*, 261-274.

Tilman, D., Cassman, K.G., Matson, P.A., Naylor, R. and Polasky, S. (2002). Agricultural sustainability and intensive production practices. *Nature*, *418*, 671-677.

Trouvelot, S., Varnier, A. L., Allègre, M., Mercier, L., Baillieul, F., Arnould, C., Gianinazzi-Pearson, V., Klarzynski, O., Joubert, J. M., Pugin, A. and Daire, X. (2008). A beta-1,3 glucan sulfate induces resistance in grapevine against *Plasmopara viticola* through priming of defense responses, including HR-like cell death. *Mol. Plant Microbe Interact.*, *21*, 232-243.

U.S. Department of Agriculture (USDA) (2010). Mexico: Greenhouse and shade house production to continue increasing. URL: http://gain.fas.usda.gov/Recent%20GAIN%20Publications/Greenhouse%20and%20Shade%20House%20Production%20to%20Continue%20Increasing_Mexico_Mexico_4-22-2010.pdf.

U.S. Environmental Protection Agency (U.S. EPA) (2012). Guidelines for water reuse. URL: http://nepis.epa.gov/Adobe/PDF/P100FS7K.pdf.

United Nations (UN). (2009). Global drylands: A UN system-wide response. URL: http://www.unccd.int/Lists/SiteDocumentLibrary/Publications/Global_Drylands_Full_Report.pdf.

Vattem, D.A. and Shetty, K. (2002). Solid-state production of phenolic antioxidants from cranberry pomace by *Rhizopus oligosporus*. *Food Biotechnol.*, *16*, 189-210.

Vera, J., Castro, J., Gonzalez, A. and Moenne, A. (2011). Seaweed polysaccharides and derived oligosaccharides stimulate defense responses and protection against pathogens in plants. *Mar. Drugs*, *9*, 2514-2525.

Wang, X., Chang, V.W. and Tang, C.Y. (2016). Osmotic membrane bioreactor (OMBR) technology for wastewater treatment and reclamation: Advances, challenges, and prospects for the future. *J. Membr. Sci.*, *504*, 113-132.

Wu, S., Shen, Z., Yang, C., Zhou, Y., Li, X., Zeng, G., Ai, S. and He, H. (2017). Effects of C/N ratio and bulking agent on speciation of Zn and Cu and enzymatic activity during pig manure composting. *Int. Biodeterior. Biodegrad.*, *119*, 429-436.

Xu, Y., Chang, Y., Chen, G. and Lin, H. (2016). The research on LED supplementary lighting system for plants. *Opt. Int. J. Light Electron Opt.*, *127*, 7193-7201.

Zodape, S.T. (2001). Seaweeds as a biofertilizer. *J. Sci. Ind. Res.*, *60*, 378-382.

Reviewed by the Oxford Science Editing.

In: Advances in Hydroponics Research
Editor: Devin J. Webster

ISBN: 978-1-53612-131-5
© 2017 Nova Science Publishers, Inc.

Chapter 7

HUMAN URINE ASSOCIATED WITH CASSAVA WASTEWATER AS A NUTRITIVE SOLUTION WITH POTENTIAL TO AGRICULTURAL USE

Narcísio Cabral de Araújo[1], Abílio José Procópio Queiroz[2], Rui de Oliveira[3], Monica de Amorim Coura[4], Josué da Silva Buriti[5] and Andygley Fernandes Mota[6]

[1]Environmental and Sanitary Engineer (UEPB), Master in Civil and Environmental Engineering (UFCG) and Doctorate Student in Agricultural Engineering by Federal University of Campina Grande (UFCG), Campina Grande, Paraiba, Brazil

[2]Environmental and Sanitary Engineer (UEPB), Master in Science and Engineering of Materials (UFCG) and Doctorate Student in Science and Engineering of Materials by Federal University of Campina Grande (UFCG), Campina Grande, Paraiba, Brazil

[3]Civil Engineer, Master and PhD in Civil Engineering, Professor Doctor C of Department of Environmental and Sanitary Engineering of the State University of the Paraiba (UEPB), Campina Grande, Paraiba, Brazil

[4]Chemistry, Master in Civil Engineering, Doctor in Natural Resources, Professor Associated IV of the Academic Unity of Civil Engineering of

the Federal University of Campina Grande (UFCG), Campina Grande, Paraiba, Brazil
[5]Industrial Chemistry (UEPB), Master in Environmental Sciences and Technology (UEPB) and Doctor in Science and Engineering of Materials by Federal University of Campina Grande (UFCG), Campina Grande, Paraiba, Brazil
[6]Agronomist Engineer, Master in Management and Conservation of Soil by the Federal University of Semi-arid (UFERSA) and Doctorate Student of Agricultural Engineering by Federal University of Campina Grande (UFCG), Campina Grande, Paraiba, Brazil

ABSTRACT

Agricultural use of wastewater containing essential nutrients for plants is an alternative to minimize environmental pollution and overexploitation of mineral fertilizers used in agriculture. In this context, this work aims to present and discuss results of physicochemical characterization of Human urine, Cassava Wastewater and alternative nutrient solution prepared through these effluents as an alternative for its use in agroecological systems of agricultural cultivation. The treatment and physicochemical characterization of the wastewater was carried out at the Federal University of Campina Grande. Following standardized methodology was characterized the hidrogenionic potential (pH), electrical conductivity (CE), chemical demand of oxygen (CDO), chloride (Cl$^-$), total phosphorus (Pt), orthophosphate (P-PO$_4^{-3}$), ammoniacal nitrogen (P-PO$_4^{-3}$), N-NH^{4+}), total nitrogen Kjeldhal (TNK), potassium (K), sodium (Na) and thermotolerant coliforms. After analyzing and comparing the results it was concluded that Human urine and Cassava Wastewater presents significant quantities of nutrients and can be used in family farming as an alternative source of fertilizers; Solutions of Human urine associated with the Cassava Wastewater present potential to be used in an agroecological culture system, aiming at the recycling of nutrients as a sustainable alternative to minimize the environmental impacts caused by the inappropriate practices of wastewater discharge into the environment.

Keywords: eco sanitation, Human urine, flour houses, alternative nutritional solution

INTRODUCTION

One of the challenges of sanitation is to minimize the environmental impacts caused by high water consumption in anthropogenic activities and the inadequate release of human excreta into the environment. Therefore, in the last years, the concept of ecological sanitation or eco sanitation has appeared "where the rationalization of the water consumption and the segregation of the effluents are applied to enable its reuse, close to the generating sources" (Silva et al., 2007). The main objective of the eco sanitation is to recycle the nutrients that are wasted in the final disposal of biodegradable waste (wastewater, biodegradable solid waste, human and animal excreta) generated in anthropic activities.

In order to effectively manage the reuse of domestic wastewater, it is customary to divide these waters into three types: Black, Yellow and Gray. Black waters: wastewater from sanitary vessels, basically containing feces, urine and toilet paper or from stool and urine separating devices, having in their composition large amounts of fecal material and toilet paper (Gonçalves et al., 2006). Gray waters are those from wash basins, showers, tanks and washing machines and dishwashers (Fiori et al., 2006). Are called of yellow waters the wastewater generated in urinals or in sanitary vessels with separating compartments for collection of urine (Costanzi et al., 2010).

In the last ten years, studies based on the separation of urine and feces have shown new development concepts for sanitation, reducing the waste of drinking water in the bathrooms and showing a new eco-friendly and economical design (Sousa et al., 2008). Urine separation is an alternative technology or even a sanitation complement that has been implemented in many places in several countries around the world (Kvarnström et al., 2006).

According to Schönning and Stenström (2004), the separation of urine brings practical and hygienic benefits, allowing its use as fertilizer and reducing the environmental effects of nutrients released by toilets, such as eutrophication of water bodies. The high nutrient load and the low content of pathogens and metals make the separation and utilization of urine a promising alternative for its use as a fertilizer (Cohim et al., 2008). According to the authors, because urine has a reduced pathogenic load, its use in small systems does not require an advanced treatment, but for large-scale reuse it is necessary to have some form of treatment, the most used being storage.

The cassava root processing agroindustry's are a source of employment and income for the rural producer. In the process of root processing, it generates a quantity of very significant residues, among which the Cassava Wastewater, which is the constituent liquid of cassava roots. According to Araújo et al. (2012) Cassava Wastewater is extracted in the stage of mass pressing processing from grated cassava roots for the production of flour and/or starch extraction. It has a milky appearance, light yellow color and foul odor, which can cause unpleasant sensations if the individual inhales for a long time at the time of its extraction.

According to Marini and Marinho (2011) the Cassava Wastewater presents potential of use for fertilization of plants in organic cultivation, since it contains macro and micronutrients. The chemical composition of Cassava Wastewater supports the potential of the compound as a fertilizer, given its richness in potassium, nitrogen, magnesium, phosphorus, calcium, and sulfur, as well as iron and micronutrients in general (Pantaroto, Cereda, 2001). Fioretto (2001) corroborates that the Cassava Wastewater can be used as fertilizer, in order to harness and recirculate the nutrients in the soil, avoiding the discharge into the watercourses.

In this context, the present study aims to present and discuss results of physical and chemical characterization of alternative nutrient solution prepared with Human urine associated with Cassava Wastewater as an alternative for the recycling of nutrients in agroecological farming systems.

MATERIAL AND METHODS

This research was developed at the Federal University of Campina Grande (UFCG), located in the city of Campina Grande, state of Paraiba, northeastern Brazil, whose geographical coordinates are 7°13'50" from Latitude (S), 35°52'52" of Longitude (W) and 551 m of altitude.

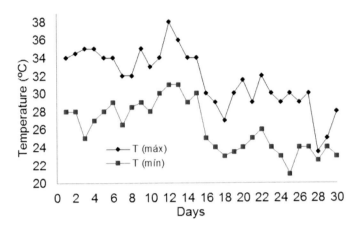

Figure 1. Maximum and minimum temperatures (°C) registers during the urine storage period.

Human urine was collected during one week in four households in Campina Grande, one in the Bodocongó neighborhood and three in the District of São José da Mata.

After collection, the urine was transported to a greenhouse located in Campus I of the UFCG, where it was treated through storage in a hermetically sealed plastic container for a period of one month. Figure 1 illustrates the daily maximum and minimum temperatures monitored during the storage period.

The used Cassava Wastewater was collected in a flour house installed in Jenipapo District, Puxinanã Municipality, PB and stored for a period of one week in a plastic container with a capacity of 20 L, which was kept with the lid partially closed, favoring output of the hourly gases generated in the Cassava Wastewater fermentation.

In the preparation of the alternative nutritional solutions, composed of Human urine associated with Cassava Wastewater, Human urine was diluted of 1, 2, 3, 4, 5% in supply water and adjusted pH to 6.4 by addition of Cassava Wastewater, because according to the United Nations Food and Agriculture Organization (FAO, 2001) and the Agronomic Institute of

Campinas (Furlani et al., 2009) nutrient solutions to be used in the cultivation of hydroponic forage and hardwood vegetables must have pH Between 6.0 and 6.5.

To adjust the pH of nutrient solutions was used a phmetro pre-calibrated. The pH of Human urine dilutions and the volumes of Cassava Wastewater used are shown in Table 1.

After adding Cassava Wastewater to adjust the pH, the dilutions of urine were denominated Nutritive solutions: S_1 (Nutritive solution contains 1% of Human urine plus Cassava Wastewater), S_2 (Nutritive solution contains 2% of Human urine plus Cassava Wastewater), S_3 (Nutritive solution contains 3% of Human urine plus Cassava Wastewater), S_4 (Nutritive solution contains 4% of Human urine plus Cassava Wastewater) and S_5 (Nutritive solution contains 5% of Human urine plus Cassava Wastewater).

The physicochemical and microbiological characterization of Human urine, Cassava Wastewater and nutrient solutions were carried out in the sanitation laboratory of the Federal University of Campina Grande.

Table 1. Cassava Wastewater volumes used in the pH correction per liter of Human urine dilutions

Human urine dilutions (%)	pH	VCW L^{-1} (mL L^{-1})
1	8.7	20.0
2	8.8	35.0
3	8.8	45.0
4	8.9	67.5
5	8.9	87.5

pH: pH of urine dilutions before of the Cassava Wastewater addition; VCW L^{-1} (mL L^{-1}): Volume of Cassava Wastewater per liter of diluted urine.

After the treatment period, the physicochemical and microbiological characterization of Human urine was determined by the following parameters: pH, electrical conductivity (EC), chemical oxygen demand (COD), chloride (Cl^-), phosphorus total (Pt), orthophosphate (P-PO_4^{-3}), ammoniacal nitrogen (N-NH$_4^+$), total nitrogen Kjeldhal (TNK), potassium (K), sodium (Na) and thermotolerant coliforms (CTT). For the Cassava Wastewater and nutrient solutions only the physicochemical parameters were determined. The analyzes followed the "Standard Methods for the Examination of Water and Wastewater" (Apha, 2005).

Table 2. Results of characterization of the Human urine and Cassava Wastewater

Ef.	pH	CE	CDO	Cl⁻	NTK	NH$_4^+$	Pt	P-PO$_4^+$	K	Na	CTT
-		mS cm^{-1}			mg L^{-1}						UFC 100 mL^{-1}
Ur.	9.0	42.6	20636	6103	6889	5760	404	393	202	675	Ausente
CWW	4.2	7.7	72290	761	968	218	420	251	475	98.5	-

Ef.: effluent; Ur.: Human urine; CWW.: Cassava Wastewater; pH: hidrogenionic potential; CE: electrical conductivity; CDO: chemical oxygen demand; Pt: phosphorus total; P-PO$_4^+$: orthophosphate; K: Potassium; Na: Sodium; TNK: Total nitrogen Kjeldahl; N-NH$_4^+$: Ammonia nitrogen; Cl⁻: chloride and CTT: thermotolerant coliforms.

RESULTS AND DISCUSSION

Table 2 shows the mean values of the physicochemical and microbiological characterization of Human urine, after 30 days of storage and Cassava Wastewater.

According to the data, in Human urine presented hydrogenionic potential (pH) of 9.0; Electrical conductivity (EC) of 42.63 mS cm^{-1}; Chemical oxygen demand (CDO) of 20636 mgO$_2$ L^{-1}; chloride (Cl⁻) of 6103 mgCl⁻L^{-1}; total nitrogen Kjeldhal (TNK) of 6889 mg N L^{-1}; Ammonia nitrogen (N-NH$_3$) of 5760 mg N-NH$_3$ L^{-1}; total phosphorus (Pt) of 404 mgP

L^{-1}; Orthophosphate (P-PO4-3) of 393 mg $P-PO_4^{-3}$ L^{-1}; Potassium (K) of 202 mg K L^{-1}; Sodium (Na) of 675 mg In L^{-1} and absence of thermotolerant coliforms (CTT).

A possible explanation for the absence of thermotolerant coliforms in Human urine is the effect of the high hydrogenation potential (pH), because under these conditions the pathogenic microorganisms are inactivated, causing the collection of urine separate from the feces, followed by treatment through temperature above 28°C, does not present a microbiological risk for its agricultural use.

The results of the physicochemical and microbiological characterizations of Human urine, after 30 days of storage (Table 2), are very close to those referenced, since when characterizing Human urine Zancheta (2007) found 7435 mg N L^{-1} of NTK; 407 mg P L^{-1} total phosphorus; 1662 mg K L^{-1} potassium; 7896 mg O_2 L^{-1} of CDO; 6000 mg $Cl-L^{-1}$ chloride; electrical conductivity of 49.0 mS cm^{-1}; pH of 9.0 and absence of thermotolerant coliforms after 21 days of storage. Rios (2008) found concentrations of 5300 mg N L^{-1} of NTK; 3500 mg $N-NH_3$ L^{-1} of ammonia; 300 mg P L^{-1} total phosphorus; 9300 mg $Cl-L^{-1}$ chloride; 6300 mg O_2 L^{-1} CDO; 44.44 mS cm^{-1} of EC; PH of 8.73 and absence of thermotolerant coliforms. Costanzi et al. (2010) found mean values of 9.32 for pH; 71.775 mS cm^{-1} of CE; 1763.8 $mgN-NH_3$ L^{-1} of ammoniacal nitrogen; 4132.2 mg N L^{-1} of NTK and 756.6 mg K L^{-1} of potassium. When characterizing Human urine stored for a period of 20 days, Botto (2013) found a pH of 9.7; EC of 42 mS cm^{-1}; absence of thermotolerant coliforms; Ammoniacal nitrogen of 5530 mg N - NH_3 L^{-1} and total phosphorus of 275 mg P L^{-1}.

As shown in Table 2, the hydrogenionic potential (pH) of Cassava Wastewater was 4.17 pH units; 7.68 mS cm^{-1} of electrical conductivity (EC); 72290 mg O_2 L^{-1} chemical oxygen demand; 761 mg $Cl-L^{-1}$ chloride; 968 mg N L^{-1} of total nitrogen Kjeldhal; 218 mg $N-NH_3$ L^{-1} of ammoniacal nitrogen; 420 mg P L^{-1} total phosphorus; 251 mg $P-PO_4^{-3}$ L^{-1} of orthophosphate; 475 mg K L^{-1} potassium and 98.5 mg sodium L^{-1}.

Duarte et al. (2012), evaluating the use of different doses of Cassava Wastewater in the lettuce crop instead of mineral fertilization, found pH in

Cassava Wastewater of 4.08; Nitrogen of 980 mg L^{-1}; 740 mg L^{-1} phosphorus; potassium of 1970 mg L^{-1} and sodium of 460 mg L^{-1}. Naples (2012) studying Cassava Wastewater and cow urine as biofertilizer in jatropha culture, when analyzing Cassava Wastewater found pH of 4.5; EC of 8.54 mS cm^{-1}; CDO of 141036 mg O_2 L^{-1}; NTK of 2049.60 mg N L^{-1} and total phosphorus of 273.12 mg P L^{-1}.

Table 3. Results of the medium values of the physicochemical characterization of the dilutions of urine associated with Cassava Wastewater

Nutritive solutions	Parameters								
	NTK	N-NH$_3$	Pt	P-PO$_4^{-3}$	K	Cl$^-$	Na	pH$_{da}$ pH$_{dd}$	CE
	mg L^{-1}							-	mS cm^{-1}
S$_1$ (1%)	85,3	60	13	10	2,6	304	15	8,7 6,4	2,17
S$_2$ (2%)	157	125	21	15	5,2	396	17	8,8 6,4	2,44
S$_3$ (3%)	234	213	31	20,5	6,8	498	18	8,8 6,4	3,0
S$_4$ (4%)	287	288	44	27,5	10,3	556	20	8,9 6,4	3,7
S$_5$ (5%)	396	360	55	30,4	13,5	639	22	8,9 6,4	4,3

Legend: NTK: Total Nitrogen Kjeldhal; N-NH$_3$: Ammoniacal nitrogen; PT: phosphorus total; P-PO$_4^{-3}$: Orthophosphate; K: Potassium; Cl$^-$: Chloride; Na: Sodium; pH$_{da}$: hydrogenionic potential of the dilution of urine before adding Cassava Wastewater; pH$_{dd}$: hydrogenionic potential of the dilution of urine after adding Cassava Wastewater e CE: electrical conductivity.

The results of physicochemical characterization of the Nutritive solutions compounds with Human urine associated with Cassava Wastewater are showed in Table 3.

According to Prado (2017) pH values of a Nutritive solution below 4 affect the integrity of the membranes (H$^+$ affects the cells, the permeability of the membranes), being able to lose already absorbed nutrients and also affects the availability of cations and pH higher than 6.5 promotes the reduction of the availability of the macronutrients Ca and P and of the micronutrients Mn, Cu, Zn and B, by the formation of precipitates, besides reducing the transport of the nutrient to the interior of the cells.

The electrical conductivity (EC) of a solution is the quantitative numerical expression of its capacity to carry the electric current due to the concentrations of salts present in the medium. According to Silva et al. (2012) high concentrations of soluble salts can seriously affect the development and production of many crops because as the concentration of salts increases in the nutrient solution the osmotic potential reduces, thus requiring greater energy from the plant to absorb water, to compromise the development of the plant due to water stress.

The nutritional requirements of any plant are determined by the amount of nutrients it extracts during its cycle. The extraction of nitrogen, phosphorus, potassium, calcium and magnesium increases linearly with the increase in production (Coelho & França, 2017).

Nitrogen is one of the macronutrients very important for plants, because according to Camargo and Silva (1987), it is an integral part of proteins, chlorophyll and enzymes. Its application in adequate amounts can favor the growth of the root, due to the fact that the growth of the aerial part increases the leaf area and the photosynthesis, and with this, a greater flow of carbohydrates to the root, favoring its growth (Prado, 2014).

Phosphorus is crucial in plant metabolism, playing an important role in cell energy transfer, respiration, and photosynthesis. It is also a structural component of the nucleic acids of genes and chromosomes, as well as of many coenzymes, phosphoproteins and phospholipids (Grant et al., 2001).

In plants, potassium deficiency decreases photosynthesis and increases respiration, reducing carbohydrate supply, and consequently plant growth (Camargo & Silva, 1987).

Using the nutrient solutions presented in Table 3 in research, Araújo et al. (2014) concluded that the use of Human urine associated with Cassava Wastewater proved to be effective for the cultivation of hydroponic green forage of maize in sugarcane bagasse substrate, since there was a statistically significant difference between the aerial part variables with gain of dry mass in all the treatments, when compared with the control.

And that Human urine plus Cassava Wastewater can replace the nutrient solution used for the cultivation of hydroponic green corn forage.

Currently, researchers from the Federal University of Campina Grande (UFCG) are developing research using Human urine and Cassava Wastewater as an alternative source of nutrients for lettuce (*Lactuca sativa* L.), pepper (*Capsicum annuum* L.), corn (*Zea mays* L.), cotton (*Gossypium hirsutum* L.) and vigna beans (*Vigna unguiculada* L.).

CONCLUSION

Human urine and Cassava Wastewater have significant amounts of nutrients and can be used in family farms as an alternative fertilizer source.

Human urine solutions associated with Cassava Wastewater have the potential to be used as an alternative source of nutrients in an agroecological system.

The recycling of nutrients contained in Cassava Wastewater and Human urine is a sustainable alternative to minimize the environmental impacts caused by improper practices of wastewater discharge into the environment.

REFERENCES

APHA - American Public Health Association; AWWA - American Water Works Association; WEF - Water Environment Federation. *Standard Methods for the examination of water and wastewater*. 21st ed. Washington DC: APHA, 2005.

Araújo, N. C. et al. Avaliação do uso de efluente de casas de farinha como fertilizante foliar na cultura do milho (*Zea mays* L.) [Evaluation of the use of effluent from flour houses as foliar fertilizer in maize (Zea mays L.)]. *Engenharia na Agricultura*, v. 20, n. 4, p. 340-349, 2012.

Araújo, N. C. et al. Produtividade de forragem hidropônica de milho (*Zea mays* l.) fertirrigado com urina e Cassava Wastewater. 12º Congresso da Água/16° Encontro de Engenharia Sanitária e Ambiental (ENASB)/XVI Simpósio Luso-brasileiro de Engenharia Sanitária e Ambiental (SILUBESA) [Hydroponic forage yield of corn (Zea mays L.) fertirrigado with urine and Cassava Wastewater. 12th Water Congress/16th Meeting of Sanitary and Environmental Engineering (ENASB)/XVI Luso-Brazilian Symposium on Sanitary and Environmental Engineering (SILUBESA)]. Lisboa, Portugal, 2014, p. 1-8.

Botto, M. P. Utilização da Urina Humana como Biofertilizante para Produção de Alimentos e Energia: Caracterização, uso na agricultura e recuperação de nutrientes. 2013. 270 f. Tese (Doutorado em Eng. Civil) - Universidade Federal do Ceará, Departamento de Engenharia Hidráulica e Ambiental [Utilization of Human urine as a Biofertilizer for Food and Energy Production: Characterization, use in agriculture and nutrient recovery. 2013. 270 f. Thesis (Doctorate in Civil Engineering) - Federal University of Ceará, Department of Hydraulic and Environmental Engineering], Fortaleza - CE, 2013.

Camargo, N. P.; Silva, O. *Manual de adubação foliar* [Foliar fertilization manual]. Editoras, La Libreria e Herba Ltda. São Paulo, SP, 1987.

Cohim, E. et al. Avaliação da Perda de Nitrogênio em Sistema de Armazenamento de Urina com Isolamento da Atmosfera. XXXI Congreso Interamericano AIDIS [Evaluation of Nitrogen Loss in Urine Storage System with Atmospheric Isolation. XXXI Inter-American Congress AIDIS]. Santiago - Chile, p. 1-8, 2008.

Costanzi, R. N. et al. Reúso de Água Amarela. *Revista de Engenharia e Tecnologia* [Reuse of Yellow Water. *Journal of Engineering and Technology*]. v. 2, n. 1, P. 9-16, 2010.

Coelho, A. M.; França, G. E. Nutrição e Adubação do Milho [Corn Nutrition and Fertilization]. Disponível em [Abailable at]: <file:///C:/Users/ Narciso/Downloads/nutricao_adubacao_milho.pdf>. Acesso em: 26 de fevereiro de 2017.

Duarte, A. S. et al. Uso de diferentes doses de Cassava Wastewater na cultura da alface em substituição à adubação mineral. *Revista Brasileira de Engenharia Agrícola e Ambiental* [Use of different doses of Cassava Wastewater in lettuce crop to replace mineral fertilization. *Brazilian Journal of Agricultural and Environmental Engineering*]. v. 16, n. 3, p. 262-267, 2012. Disponível em: http://www.agriambi.com.br. Acesso em 26 de fevereiro de 2017.

FAO - Organización de la Naciones Unidas Para la Agricultura Y Alimentación. Forraje Verde Hidropónico. Oficina Regional de la FAO para América Latina y el Caribe. *Manual Técnico*. Primera Parte [United Nations Organization for Agriculture and Food. Hydroponic Green Fodder. FAO Regional Office for Latin America and the Caribbean. *Technical manual*. First part], 2001, 68p.

Fioretto, R. A. Uso Direto da Cassava Wastewater em Fertirrigação [Direct Use of Cassava Wastewater in Fertigation]. In: Cereda, M. P (Coord.): Manejo, Uso e tratamento de subprodutos da industrialização da mandioca. Fundação Cargill [Handling, use and treatment of by-products of cassava industrialization. Cargill Foundation], v. 4, p. 67-79. São Paulo, SP, 2001.

Fiori, S.; Fernandes, V. M. C.; Pizzo, H. Avaliação Qualitativa e Quantitativa do Reúso de Águas Cinzas em Edificações [Qualitative and Quantitative Assessment of Gray Water Reuse in Buildings]. Ambiente Construído, Porto Alegre, v. 6, n. 1, p. 19-30, 2006.

Furlani, P. R. et al. Cultivo hidropônico de plantas: parte 2 - Solução Nutritiva. 2009. Artigo em Hypertexto [Hydroponic plant cultivation: part 2 - Nutritive solution. 2009. Article in Hypertext]. Disponível em: <http://www.infobibos.com/Artigos/2009_2/hidroponiap2/index.htm>. Acesso em: 26 de fevereiro de 2017.

Gonçalves, R. F. et al. Caracterização e Tratamento de Diferentes Tipos de Águas Residuárias de Origem Residencial Após Segregação. AIDIS - Asociación Interamericana de Ingeniería Sanitaria y Ambiental. Sección Uruguay. Rescatando Antiguos Principios para os Nuevos Desafíos del Milenio [Characterization and Treatment of Different Types of Residual Waters of Residential Origin after Segregation. AIDIS - Inter-American Association of Sanitary and Environmental Engineering. Uruguay Section. Rescuing Old Principles for the New Millennium Challenges]. Montevideo, p. 1-10, 2006.

Grant, C. A. et al. A Importância do Fósforo no Desenvolvimento Inicial da Planta. POTAFOS - Associação Brasileira para Pesquisa da Potassa e do Fosfato. Informações Agronômicas [The Importance of Phosphorus in the Initial Development of the Plant. POTAFOS - Brazilian Association for Potash and Phosphate Research. Agronomic Information], n° 95, Piracicaba-SP, 2001.

Marini, F. S.; Marinho, C. S. Adubação Complementar para a Mexeriqueira 'Rio' em Sistema de Cultivo Orgânico. Revista Brasileira de Engenharia Agrícola e Ambiental, Campina Grande, PB [Complementary Fertilization for the 'Rio' Mexceriqueira in Organic Cultivation System. *Brazilian Journal of Agricultural and Environmental Engineering*, Campina Grande, PB]. v. 15, n. 6, p. 562-568, 2011. Disponível em: <http://www.agriambi.com.br>. Acesso 26 de fevereiro de 2017.

Nápoles, F. A. M. Tecnologia Agroecológica de Cultivo do Pinhão-manso Utilizando Urina de Vaca e Cassava Wastewater. 2012. 127 f. Tese (Doutorado em Engenharia Agrícola) - Universidade Federal de Campina Grande, Centro de Tecnologia e Recursos Naturais [Agroecological Technology of Jatropha Cultivation Utilizing Cow Urine and Cassava Wastewater. 2012. 127 f. Thesis (PhD in Agricultural Engineering) - Federal University of Campina Grande, Center for Technology and Natural Resources], Campina Grande - PB, 2012.

Pantaroto, S.; Cereda, M. P. Linamarina e sua decomposição no ambiente. In: Cereda, M.P (coord.): *Manejo, uso e tratamento de subprodutos da industrialização da mandioca* [Linamarin and its decomposition into the environment. In: Cereda, M.P (coord.): *Management, use and treatment of by-products of manioc industrialization*]. Fundação Cargill [Cargill Foundation], v. 4, p. 38-47. São Paulo, SP, 2001.

Zancheta, P. G. Recuperação e Tratamento da Urina Humana Para Uso Agrícola. 2007, 83 f. Dissertação (mestrado) - Universidade Federal do Espírito Santo, Centro Tecnológico [Recovery and Treatment of Human urine for Agricultural Use. 2007, 83 f. Dissertation (master's degree) - Federal University of Espírito Santo, Technological Center], Vitória - ES, 2007.

Prado, R. M. Manual de Nutrição de Plantas Forrageiras [Manual of Nutrition of Forage Plants]. Disponível em: <http://www.nutricaode plantas.agr.br/site/downloads/sumula_livro_nutricaoforrageira.pdf>. Acesso em 16 de fevereiro de 2017.

Rios, É. C. S. V. Uso de Águas Amarelas Como Fonte Alternativa de Nutriente em Cultivo Hidropônico da Alface (Lactuca sativa). 2008. 105 f. Dissertacão (Mestrado em Eng. Ambiental) - Universidade Federal do Espirito Santo [Use of Yellow Waters as an Alternative Source of Nutrient in Hydroponic Lettuce Cultivation (Lactuca sativa). 2008. 105 f. Dissertacao (Master in Environmental Engineering) - Federal University of Espirito Santo], Vitória - ES, 2008.

Schönning, C.; Stenström, T. A. Diretrizes para o Uso Seguro de Urina e Fezes nos Sistemas de Saneamento Ecológico. Instituto Sueco de Controle de Doenças Infecciosas [Guidelines for the Safe Use of Urine and Stool in Ecological Sanitation Systems. Swedish Institute for Infectious Disease Control]. Programa EcoSanRes, Instituto Ambiental de Estocolmo, Estocolmo, Suêcia, 2004, p. 38.

Sousa, J. T. et al. Gerenciamento sustentável de água residuária doméstica [Sustainable management of domestic wastewater]. *Revista Saúde e Ambiente* [*Health and Environment Journal*], v. 9, n. 1, 2008.

Silva, A. B. et al. Avaliação do desenvolvimento inicial da Helicônia *bihai* em substrato inerte irrigado com diferentes níveis de diluição de urina humana em casa de vegetação. Conferência Internacional em Saneamento Sustentável: Segurança alimentar e hídrica para a América Latina [Evaluation of the initial development of bihai Heliconia in irrigated inert substrate with different levels of dilution of Human urine in greenhouse. International Conference on Sustainable Sanitation: Food Security and Water for Latin America]. Ecosan, Fortaleza, CE, 2007.

Silva, J. B. G. et al. Avaliação da Condutividade Elétrica e pH da Solução do Solo em Uma Área Fertirrigada com Água Residuária de Bovinocultura de Leite [Evaluation of the Electrical Conductivity and pH of the Soil Solution in a Fertigated Area with Milk Cattle Residual Water]. Irriga, Botucatu, Edição Especial, p. 250-263, 2012.

Kvarnström et al. Separação de Urina: Um Passo em Direção ao Saneamento Sustentável. Programa EcoSanRes. Instituto Ambiental de Estocolmo, Estocolmo, Suécia, 2006 [Urine Separation: A Step Toward Sustainable Sanitation. EcoSanRes Program. Stockholm Environmental Institute, Stockholm, Sweden, 2006]. Disponivel em: <www.ecosanres. org>. Acesso em 10 de fevereiro de 2017.

In: Advances in Hydroponics Research
Editor: Devin J. Webster

ISBN: 978-1-53612-131-5
© 2017 Nova Science Publishers, Inc.

Chapter 8

APPLICATION OF HYDROPONIC CULTURE FOR THE CULTIVATION OF EDIBLE CACTI

Takanori Horibe[*]

College of Bioscience and Biotechnology, Chubu University,
Kasugai, Aichi, Japan

ABSTRACT

Hydroponic culture is a method of growing plants using nutrient solution (water and fertilizer) with or without the use of an artificial medium. The lack of soil means absence of weeds or soil-borne disease, thus making precise fertilizer management possible. Thus, hydroponic culture conveys numerous advantages for edible *Opuntia* production, which is conventionally produced through soil or pot culture. To date, we have investigated the effects of hydroponic culture involving the deep flow technique on the growth of edible *Opuntia* and showed that can be grown effectively using hydroponics compared with the commercially practiced pot culture. Thus, hydroponics may be an ideal method of edible *Opuntia* cultivation for farmers and horticultural hobbyists based in cities, who typically practice soil and pot culture. In this review, we describe recent progress and findings on the hydroponic culture of edible *Opuntia*.

---

[*] Corresponding Author Email: t-horibe@isc.chubu.ac.jp, Fax: +81-092-642-2913.

INTRODUCTION

The Cactaceae are an exciting group of plants because of their varied morphology, succulence, and their adaptations to the environment. This family includes over 1500 species belonging to ca. 127 genera (Barthlott and Hunt, 1993; Hunt et al., 2006). Cacti occur naturally from just south of the Arctic Circle in Canada to the tip of Patagonia in South America (Rebman and Pinkava. 2001), and its greatest species richness is concentrated primarily in Mexico. In addition, cacti show great adaptation to various environments. They grow at altitudes from below sea level to over 4,500 m in the Andes; and in climates having little rainfall to more than 500 cm of annual precipitation. Size of cacti also vary from 3 cm to as tall as 20+ m and weighing several tons (Rebman and Pinkava. 2001).

Subfamily *Pereskioideae*, *Opuntioideae*, and *Cactoideae* have been recognized as distinct subfamilies within Cactaceae from taxonomic studies since the 19th century (Anderson, 2001; Metzing and Kiesling, 2008). The genus *Maihuenia* has been typically considered as a member of *Pereskioideae*. However, placement of *Maihuenia* in a monogeneric subfamily has been suggested on the basis of its unique ecological and morphological attributes (Anderson, 2001) and molecular phylogenetic analyses (Wallace, 1995a, b). Nyffeler (2002) suggested that species of *Pereskia* and *Maihuenia* were found to form an early-diverging grade within Cactaceae, with *Cactoideae* and *Opuntioideae* as well-supported clades. Members of subfamily *Pereskioideae* are large trees or shrubs with thin, broad, ordinary-looking leaves and hard, woody, non-succulent trunks: are not adapted to dry, hot conditions (Mauseth, 2006). Subfamilies *Maihuenioideae* and *Opuntioideae* contain plants with small but still easily visible foliage leaves, and plants vary from being trees to dwarfs (Mauseth, 2006). The largest subfamily, *Cactoideae*, are characterized by tubercles or ribs on the stems, reduced or suppressed leaves subtending each areole (Wallace and Gibson, 2002).

The flat-stemmed prickly-pear cactus is a crop with a high capacity of adaptation to different environmental conditions, mainly arid and semiarid climates. It belongs to the genus *Opuntia* (subfamily *Opuntioideae*, family Cactaceae), with a representation of over 100 species natives to Mexico (Bravo 1978). The species of *Opuntia* are a major source of fruit, vegetable and forage in areas, where the soils are poor or are becoming poor and result in very low yield of traditional cultivars (García-Saucedo et al., 2005). Some of these *Opuntia* species have been used by human beings for hundreds of years in various ways: for animal feed, in the production of natural dyes such as cochineal red, as vegetables for human consumption and as source of nutraceutical compounds (Guzmán-Maldonado and Paredes–López 1999; Silios-Espino et al., 2006). For example, The stems of *Opuntia* are widely consumed as a vegetable in Mexico, Latin America, South Africa, and Mediterranean countries (Stintzing and Carle 2005; Cruz-Hernández and Paredes-López 2010; El-Mostafa et al., 2014). In some countries, *Opuntia* species are also used as remedies and folk medicine for a variety of health problems including burns, edema, and indigestion (El-Mostafa et al., 2014; Shetty et al., 2012). Edible Opuntia is also produced in Japan, where they are produced mainly in Kasugai City, Aichi Prefecture, although production scale is still small. We are working with Kasugai City to promote production and consumption of edible *Opuntia* in Japan.

The *Opuntioideae* differ from all other subfamilies in having glochids (small, barbed, and deciduous spines) and seeds that are completely enwrapped by a funicular stalk, which becomes hard and bony (Rebman and Pinkava. 2001). With respect to its growth behavior, daughter cladodes develop from the areole of the mother cladode and this process is repeated (Pimienta-Barrios et al., 2005). The areole is the most distinctive morphological trait and is recognized as a Cactaceae synapomorphy. It has been traditionally recognized as a structure homologous to the axillary bud (Gibson and Nobel, 1986). However, the areole is regarded as a short shoot covered with trichomes and dynamically produces stems (long shoots), leaves, spines and/or flowers (Mauseth, 2006). The enclosed areole meristem and internal bud development are understood to be an adaptation to protect the meristem and the bud from low temperatures (Sánchez et al., 2015).

Furthermore, edible *Opuntia* exhibits crassulacean acid metabolism (CAM), a CO_2-concentrating mechanism that potentially leads to higher optimal temperatures for photosynthesis (Monson 1989). CAM plants have certain anatomical modifications that enable them to survive droughts, such as thick cuticles and low stomatal frequency, together with night-time

CO_2 uptake (Drennan and Nobel 2000; Pimienta-Barrios et al., 2005), although daughter cladodes show C3 photosynthesis with daytime stomatal opening during the early stages of development (Osmond 1978; Acevedo et al., 1983) and import water from mother cladodes (Pimienta-Barrios et al., 2005). Growth responses of *Opuntia* plants to temperature and CO_2 concentration are well investigated (Gulmon and Bloom 1979; North et al., 1995). Elevated CO_2 concentrations increase the daily net CO_2 uptake of cladodes and lead to increased biomass production (Cui et al., 1993; Nobel and Israel 1994).

Opuntia plants are commonly produced through soil or pot culture; however, the cultivation of vegetables using soil exposes them to soil-borne diseases and salt accumulation and also poses difficulties in fertilizer management (Lakkireddy et al., 2012). In a hydroponic culture, plants are grown using a nutrient solution (water and fertilizer) with or without the use of an artificial medium. Absence of soil circumvents weeds and soil-borne diseases, while precise fertilizer management is readily achieved (Lakkireddy et al., 2012). There are many advantages associated with the hydroponic culture of edible *Opuntia*, although no studies, to our knowledge, have investigated the effects of hydroponic culture on *Optunia* growth. Thus, we started investigating the effects of hydroponic culture on the growth of edible *Opuntia*.

HYDROPONIC CULTURE OF EDIBLE CACTI AND ITS EFFECT ON CLADODE DEVELOPMENT

Most cacti including edible *Opuntia* have spines and glochids on their stem surface. In Japan, edible *Opuntia* called "Maya" is also produced as a vegetable, mainly in Kasugai City, Aichi Prefecture. *Opuntia* "Maya" has fewer spines than other cultivars, but these spines are very sharp, which limits the commercial value of this variety. A number of beneficial functions have been ascribed to spines, including participation in zoochorous dispersal (Bobich and Nobel 2001; Frego and Staniforth 1985), mechanical protection from herbivores (Norman and Martin 1986), shading of the stem (Nobel et al., 1986), collection of fog to absorb water (Ju et al., 2012), reflection of light (Loik, 2008), and thus a reduction in water loss (Stintzing and Carle, 2005). However, spines on cladodes are one of the most undesirable characteristics of edible *Opuntia* for consumers. They are usually burned or removed using a knife before the cactus is put on sale.

We first investigated the effects of hydroponic culture involving the deep flow technique (DFT) on the growth of edible *Opuntia*, and its effect on spine development on daughter cladodes (Horibe and Yamada, 2016). We first attempted hydroponic (DFT) cultivation of edible *Opuntia* using commercially available bubble wrap, to see whether this was feasible. In this hydroponics culture treatments (DFT), cladodes were fixed in a container filled with deionized water using bubble wrap (Figure 1). Aeration was conducted using an air pump (Silent β-30, Nisso, Japan). Cladodes under DFT were maintained at 25°C for a 14-h light period (220—240 µmol·m^{-2}·s^{-1}), and 15°C for an 10-h dark period. Relative humidity was maintained at 70% during light and dark periods. we did not use any fertilizer in this hydroponics treatment. Daughter cladodes and roots appeared from the mother cladodes under DFT treatment (Figure 2), and the total fresh weight (FW) of cladodes also increased after treatment, showing that edible *Opuntia* can be grown by DFT. DFT is also a basic method for plant production in plant factories. Plant factories are horticulture greenhouses or automated facilities where vegetables and crops can be produced throughout the year controlling environmental conditions, such as light, temperature, humidity, CO_2, and nutrient availability by ICT (Yanata et al., 2014; Hirama, 2015). There are two main types of plant factories. These include artificial lighting-based fully enclosed systems and natural sunlight-based systems. Hydroponic culture is mainly used as the cultivation system in enclosed plant factories, whereas both soil and hydroponic technologies are used in natural sunlight systems. Plant factories using artificial light have already been used in Japan for the production of leafy vegetables. We are now trying stable production of edible *Opuntia* and increasing its nutritive value through environment control in a plant factory.

Figure 1. Hydroponic culture (DFT) of edible *Opuntia* using bubble wrap.

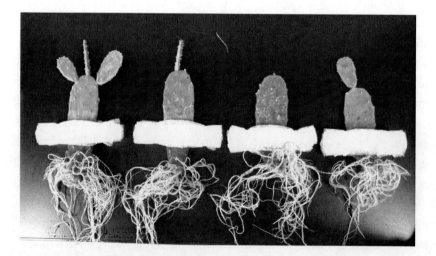

Figure 2. Cladodes of *Opuntia* 5 weeks after hydroponic culture. Daughter cladodes and roots appeared from the mother cladode.

We also evaluated the effect of drought stress on cladode growth and spine development through raising cladodes by hydroponics and pot culture with different frequencies of watering. Cladodes in pots were watered with 500 mL/pot water at the following frequencies to evaluate the effect of different levels of drought stress on the growth of daughter cladodes: once a week (low drought stress), once every 2 weeks (middle drought stress), and once every 4 weeks (high drought stress). We measured the number of areoles which had spines to calculate spine occurrence on daughter cladodes: spine occurrence = the total number of areoles which had spines longer than 1 mm/the total number of areoles. The number of daughter cladodes did not significantly differ among treatments, although length of the daughter cladodes were affected by drought stress and growth conditions (Horibe and Yamada, 2016). When given the drought stress for edible *Opuntia*, the length of first daughter cladodes in low drought stress was longest, followed by middle drought stress, high drought stress (Horibe and Yamada, 2016). Thus, drought stress seemed to suppress the elongation growth of daughter cladodes. We think that high drought stress affected the water content in cladodes and the growth of daughter cladodes, resulting in a slow growth rate of cladodes. Then, we compared spine occurrence on daughter cladodes in hydroponics and pot cultures. It was the highest in cladodes under high drought stress and the lowest under DFT (Horibe and Yamada, 2016). Among pot culture, cladodes under low and middle drought stress had fewer spines than those under high drought stress (Horibe and Yamada, 2016). This result suggests that water availability for mother cladodes affects the development of spines on daughter cladodes, showing that this approach is effective to overcome the issue of spines and their reduction of the commercial value of edible *Opuntia*. As mentioned above, Ju et al., (2012) reported that cactus spines also function as a fog collection system, in addition to other functions such as protection from herbivores. Thus, drought stress on cladodes might promote the development of spines to facilitate efficient fog collection.

Spines on cladodes are one of the most undesirable characteristics in edible *Opuntia*. Thus, cultivation technics that can reduce the number of spines can contribute to improving the commercial value of edible cacti. However, the development of spines are affected by several environmental factors including temperature, humidity, and light intensity. Thus, further research is needed to further understand the effects of hydroponics culture and cultivation conditions on the development of spines on cladodes.

EFFECTS OF LIGHT ENVIRONMENT ON CLADODE GROWTH AND SPINE DEVELOPMENT

Next, we investigated the relationship between light enviromment and cladodes growth. Light is essential for plant growth, with both its wavelength and its intensity affecting plant growth and morphogenesis (Mortensen and Stromme 1987; Yanagi et al. 2006). Studies have shown that light intensity affects the elongation growth of *Opuntia* daughter cladodes and the malate content of cladodes (Littlejohn and Ku, 1985; North et al., 1995). However, to the best of our knowledge, few studies have investigated the relationship between the light environment and the growth of edible cacti (North et al., 1995). Understanding the relationship between environmental conditions and cladode growth is important for improving the production and quality of edible cacti. Thus, we cultivated *Opuntia* using hydroponic culture and investigated the effects of red and blue LED light on its growth (Horibe et al., 2016). *Opuntia* cladodes were cultivated using a two-layer hydroponic system (Churitsu Electric Co., Japan) and cultivation panels (88 cm long × 57 cm wide × 4 cm high) in a closed-type plant factory (Figure 3). OAT House solution A (OAT Agrio Co., Ltd., Japan), with an electrical conductivity of 2 mS/cm, was used as the hydroponic nutrient solution. The light sources were red and blue LEDs (Churitsu Electric Co., Japan). Cladodes were then retained under constant light as follows: red light, peak emission at 660 nm (red); blue light, peak emission at 440 nm (blue); and simultaneous irradiation with red and blue light (red + blue), all at a photosynthetic photon flux density of 180 $\mu mol/m^2/s$. The temperature and relative humidity were maintained at 28°C and 60%–80%, respectively.

Figure 3. Cladodes planted on cultivation panel.

In all treatments, daughter cladodes developed from mother cladodes, although the width of daughter cladodes under red light was slightly less than that of daughter cladodes under other treatments (Horibe et al., 2016). The speed of elongation was the highest in cladodes under red light treatment and lowest in those under blue light treatment. Growth speed was intermediate in cladodes under simultaneous irradiation with red and blue light. The total FW of mother cladodes under simultaneous irradiation with red and blue light became higher than that of those under blue light and did not differ between other treatments (Horibe et al., 2016). The width of daughter cladodes was smallest in those treated with red light, resulting in the decrease of average FW of daughter cladodes. However, due to the large number of cladodes with red light treatment, the total FW of daughter cladodes harvested from one mother cladode did not decrease. The total FW of daughter cladodes was the smallest with blue light treatment because the number of daughter cladodes was less. Therefore, red light is more effective than blue light in increasing cladode production. Our results also suggest that *Opuntia* cladodes can undergo C3 photosynthesis via daytime stomatal opening under constant light conditions, although it is not clear whether mother cladodes also undergo C3 photosynthesis. In addition, daughter cladodes in all treatments became relatively long and narrow compared with those commonly produced in the greenhouse. Cladodes grown by DFT in a greenhouse had shapes similar to those grown in a greenhouse. Therefore, the constant light condition might have affected the elongation growth of daughter cladodes, resulting in their slender appearance. Under the blue light treatment, the speed of elongation growth of daughter cladodes became the slowest and the number of daughter cladodes was the lowest compared with those in the other treatments. Thus, compared with red light, blue light appears to suppress the development of daughter cladodes. Growth suppression by blue light has been reported in many plants (Kigel and Cosgrove 1991; Maas et al., 1995). Zhao et al,. (2007) showed that the amount of hypocotyl GA_4 decreases after irradiation with blue light, resulting in the suppression of elongation growth. Thus, changes in the concentrations of plant hormones, such as gibberellin, in response to blue light treatment might be involved in the observed suppression of growth in daughter cladodes.

Interestingly, the occurrence of spines on daughter cladodes also changed among the treatments. The number of spines on the cladodes was the highest on those treated with simultaneous irradiation with red and blue light and the lowest on those treated with red light alone. These results suggest that blue light has a stronger effect on spine development than red light. In addition, red and blue light might induce spine development via different signal transduction pathways as simultaneous irradiation with red and blue irradiation had the strongest effect on spine development. Phytochrome and photropins have been shown to interact with each other (Devlin and Kay 2000; Hughes et al., 2012); such interaction might also affect the development of spines on cladodes. Our study showed that edible cacti can be grown using LEDs in a plant factory and that the light wavelength strongly affects the growth and number of spines on daughter cladodes (Horibe et al., 2016). Manipulating the light environment to promote daughter cladode growth and suppress spine development could improve the quality and production level of edible cacti. However, other cultivation conditions, including fertilizer use, growth temperature, humidity, and light intensity, might also affect the development of cladodes and spines. Thus, further research is needed to obtain a deeper understanding of the effects of hydroponic culture and cultivation conditions on the development of spines on cladodes.

Hydroponic culture involves growing plants without soil. This system can avoid the costly and time-consuming task of soil sterilization to prevent soil-borne disease and enable precise fertilizer management (Lakkireddy et al., 2012; Wahome et al., 2011). Furthermore, it can convey numerous advantages to consumers, as well as for the cultivation of various vegetables and crops. Nitrogen is an essential element for plant growth, and nitrate is one of the available forms of nitrogen for plant uptake. Nitrate itself is relatively non-toxic, but it can be converted to harmful nitrites post-harvest, and high human intake of nitrate may produce a number of health problems (Santamaria, 2005). In hydroponics, several methods have been established to reduce the nitrate content of vegetables (Wang et al., 2007; Stefanelli et al., 2011). Ogawa (2012) reported methods for cultivating leafy vegetables and tomatoes with low potassium content for dialysis patients, in whom potassium intake is restricted. In addition, precise salinity control in hydroponics for tomato cultivation was also reported to improve fruit quality (Sakamoto et al., 1999). Although hydroponics is not commonly practiced in the cultivation of edible cacti, we think that it is also an effective way to cultivate edible cacti and improve their quality.

A COST-EFFECTIVE AND PRODUCTIVE METHOD OF HYDROPONIC CULTURE OF EDIBLE CACTUS

We have shown that edible *Opuntia* can be grown by hydroponic culture using commercially available bubble wrap and a cultivation panel. However, it was difficult to attach cladodes using these items due to their characteristic stem shape, and cladodes occasionally fell into the culture solution as the fixative loosened. For the hydroponic culture of most vegetables, seeds are spread on a commercially available cultivation panel; however, in the case of edible *Opuntia*, vegetative propagation using the stem is commonly used for its production because this method is much faster and easier than seed propagation. Therefore, the development of an appropriate method for the hydroponic culture of *Opuntia* is necessary.

Against this background, we proposed a simple and cost-effective hydroponic system that does not require a power source, which makes this approach widely applicable (Figures 4 and 5). Specifically, we used an L-shaped plastic bar (L-type angle) and clips to attach cladodes to the container during the hydroponic culture, which stabilized the cladodes and prevented any from falling into the container (Figures 4 and 5). The addition of weights on the ends of the L-shaped plastic bar may be helpful to maintain cladode stability when the cladode height and size increase.

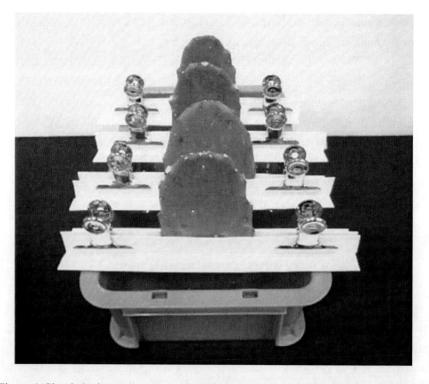

Figure 4. Simple hydroponic culture of edible *Opuntia*. L-shaped plastic bar (L-type angle) and clips were used to fix a plant; then, it was set on the plastic container.

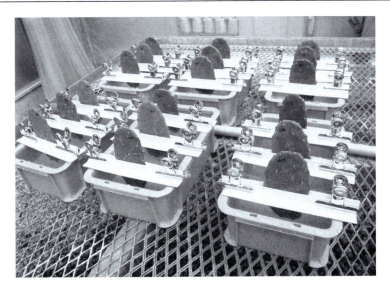

Figure 5. Simple hydroponic culture in a greenhouse.

We also investigated the effectiveness of this method by comparing the cladode growth using this method with pot culture using a growth chamber and a greenhouse. In the treatments, cladodes were attached to a plastic container using L-shaped plastic bar, and black plastic bags can be used to block light and to prevent algal growth (Figures 4 and 5). OAT House solution A (OAT Agrio Co., Ltd., Japan), with an electrical conductivity of 2 mS/cm, was used as the hydroponic nutrient solution. Cladodes grown in the growth chamber were maintained at 25 °C for a 14 h light period (160—180 µmol m^{-2} s^{-1}) and at 15 °C for a 10 h dark period. Daughter cladodes developed from the mother cladodes and kept growing in all treatments, and the growth rate became higher in hydroponic treatment than in pot culture, showing the feasibility of this new hydroponic method (unpublished results). The total FWs of cladodes harvested from a mother cladode was higher in the greenhouse as compared with that in the growth chamber in all the treatments. Environmental conditions, including photoperiod, temperature, and light intensity are reported to affect the elongation of plant stems (Shibutani and Kinoshita, 1968; Hidaka et al., 2014). Therefore, different environmental conditions including temperature and light intensity appeared to have affected the development and elongation of daughter cladodes, resulting in a higher yield in the greenhouse. Temperatures in the greenhouse during the month of August sometimes exceeded 55°C, and the water temperature of the hydroponic culture reached 42°C. However, no damage or necrosis was observed on the roots of *Opuntia* grown by hydroponic culture, even without aeration, demonstrating the surprising tolerance of *Opuntia* roots to heat and a low-oxygen environment.

By using the hydroponic system developed using low-cost materials, edible *Opuntia* cultivation can be easily performed anywhere, including in greenhouses, in buildings, as well as in areas experiencing soil stress. The hydroponic system proposed in our study should facilitate easier unfastening of cladodes from plastic bars and containers compared with that using the commercially available cultivation panels, which can be a helpful feature within a hydroponic operation for the cleaning and planting of cladodes, and for the monitoring of cladode conditions. A reduction of labor required for harvesting and management work can be achieved by positioning the hydroponic system on a high bench. In addition, this system may be ideal for use by horticultural hobbyists based in cities, who typically practice soil pot culture. Apart from its commercial application, the proposed system can also be used in research of the different *Opuntia* growth variables and their interrelationships. This includes a non-destructive method of studying the root system of *Opuntia* and any standing crop that is normally hidden below the ground. We believe that our new hydroponic cultivation method and idea of clamping cladodes for attachment to vessels, with continual improvement, can lead to the development of a hydroponic system suitable for the mass production of edible *Opuntia* in greenhouses and plant factories.

CONCLUSION

We have shown that edible cactus can be efficiently cultivated by hydroponic culture and that controlling environmental conditions can lead to improvement in its quality. These cactus plants are consumed in many countries and have various health benefits. We believe that the application of hydroponic culture for the cultivation of edible cacti can bring numerous benefits to the edible cactus industry and consumers.

REFERENCES

Acevedo, E., Badilla, I. and Nobel, P. S. (1983). Water relations, diurnal activity changes, and productivity of a cultivated cactus *Opuntia ficus-indica*. *Plant Physiol.* 72: 775-780.

Anderson, E., F. (2001). *The cactus family*. Timber Press, Portland, Oregon, USA.

Barthlott, W. and Hunt, D. R. (1993). *Cactaceae*. In K. Kubitzki [ed.], The families and genera of vascular plants. Springer-Verlag, New York, New York, USA.

Bobich, E. G. and Nobel, P. S. (2001). Vegetative reproduction as related to biomechanics, borphology and anatomy of four cholla cactus species in the Sonoran Desert. *Ann. Bot.* 87: 485-493.

Bravo H (1978) *Las Cactáceas de México*, 2nd edn. UNAM, México.

Cui, M., Miller P. S. and Nobel P. S. (1993). CO_2 Exchange and growth of the crassulacean acid metabolism plant *Opuntia ficus-indica* under elevated CO_2, in open-top chambers. *Plant Physiol.* 103: 519-524.

Cruz–Hernández, A. and Paredes–López, O. (2010). Enhancement of economical value of nopal and its fruits through biotechnology. *JPACD* 12: 110-126.

Devlin, P. F. and Kay, S. A. (2000). Cryptochromes are required for phytochrome signaling to the circadian clock but not for rhythmicity. *Plant Cell* 12: 2499-2510.

Drennan P. M. and Nobel P. S. (2000). Responses of CAM species to increasing atmospheric CO2 concentration. *Plant, Cell and Environ.* 23:767-781.

El-Mostafa, K., El-Kharrassi, Y., Badreddine, A., Andreoletti, P., Vamecq, J., El-Kebbaj, M. S., Latruffe, N., Lizard, G., Nasser, B. and Cherkaoui-Malki, M. (2014). Nopal cactus (Opuntia ficus-indica) as a source of bioactive compounds for nutrition, health and disease. *Molecules* 19:14879-14901.

Frego, K. A. and Staniforth, R. J. (1985). Factors determining the distribution of *Opuntia fragilis* in the boreal forest of southeastern Manitoba. *Can. J. Bot.* 63: 2377-2382.

García-Saucedo, P. A., Valdez-Morales, M., Valverde, M. E., Cruz-Hernández, A. and Paredes-López, O. (2005). Plant regeneration of three Opuntia genotypes used as human food. *Plant Cell Tissue Organ Cult* 80:215-219.

Gibson, A. C. and Nobel, P. S. (1986). *The cactus primer*. Cambridge: Harvard University Press.

Gulmon, S. L., Bloom A. J. (1979). C3 photosynthesis and high temperature acclimation of CAM in *Opuntia basilaris* engelm. and bigel. *Oecologia* 38: 217-222.

Guzmán-Maldonado, S. H. and Paredes–López, O. (1999). Biotechnology for the improvement of nutritional quality of food crop plants. In Paredes-López O (ed) Molecular biotechnology for plant food production. *CRC Press, Boca Raton*, 553-620.

Hidaka, K., Okamoto, A., Araki, T., Miyoshi, Y., Dan, K., Imamura, H., Kitano, M., Sameshima, K., Okimura, M. (2014). Effect of Photoperiod of Supplemental Lighting with Light-emitting Diodeson Growth and Yield of Strawberry. *Environ. Control Biol.* 52: 63-71.

Hirama, J. (2015). The history and advanced technology of plant factories. *Environ. Control Biol.* 53:47-48.

Horibe, T. and Yamada, K. (2016). Hydroponics culture of edible *Opuntia* 'Maya': drought stress affects the development of spines on daughter cladodes. *Environ. Control Biol.* 54: 153-156.

Horibe, T., Iwagaya, Y., Kondo, H. and Yamada, K. (2016). Hydroponics Culture of Edible Opuntia 'Maya': Effect of Constant Red and Blue Lights on Daughter Cladodes Growth and Spine Development. *Environ. Control Biol.* 54: 165-169.

Hughes, R. M., Vrana, J. D., Song, J. and Tucker, C. L. (2012). Light-dependent, dark-promoted interaction between *Arabidopsis* cryptochrome 1 and phytochrome B proteins. *J. Biol. Chem.* 287: 22165-22172.

Hunt, D., Taylor, N., P. and Charles, G. (2006). *The new cactus lexicon*. dh Books, Milborne Port, UK.

Ju, J., Bai, H., Zheng, Y., Zhao, T., Fang, R. and Jiang, L. A. (2012). Multi-structural and multi-functional integrated fog collection system in cactus. *Nat. Commun.* 3: 1247.

Kigel, J. and Cosgrove, D. J. (1991). Photoinhibition of stem elongation by blue and red light: effects on hydraulic and cell wall properties. *Plant Physiol.* 95: 1049-1056.

Lakkireddy, K. K. R., Kasturi K. and Sambasiva, R. K. R. S. (2012). Role of hydroponics and aeroponics in soilless culture in commercial food production. *JAST* 1: 26-35.

Littlejohn, R.O. and Ku, M. S. B. (1985). Light and temperature regulation of early morning crassulacean acid metabolism in *Opuntia erinacea var columbiana* (Griffiths) L. Benson. *Plant Physiol.* 77: 489-491.

Loik, M. E. (2008). The effect of cactus spines on light interception and Photosystem II for three sympatric species of Opuntia from the Mojave Desert. *Physiol. Plant.* 134: 87-98.

Maas, F. M., Bakx, E. J. and Morris, D. A. (1995). Photocontrol of stem elongation and dry weight portioning in Phaseolus vulgaris L. by the blue-light content of photosynthetic photon flux. *J. Plant Physiol.* 146: 665-671.

Mauseth, J. D. (2006). Structure-Function relationship in highly modified shoots of *Cactaceae*. Ann. Bot. 98:901-926.

Metzing, D. and Kiesling, R. (2008). The study of cactus evolution: The pre-DNA era. *Haseltonia* 14: 6-25.

Monson R. K. (1989). On the evolutionary pathways resulting in C_4 photosynthesis and Crassulalacean acid metabolism (CAM). *Adv. in Ecol. Res.* 19:57-110.

Mortensen L. M. and Stromme, E., (1987). Effects of light quality on some greenhouse crops. *Scientia Hort.* 33: 27-36.

Nobel, P. S., Geller, G. N., Kee, S. C. and Zimmerman, A. D. (1986). Temperatures and thermal tolerances for cacti exposed to high-temperatures near the soil surface. *Plant Cell Environ.* 9: 279-287.

Nobel, P. S., Israel, A. A. (1994). Cladode development, environmental responses of CO2 uptake, and productivity for *Opuntia ficus-indica* under elevated CO2. J. Exp. Bot. 45: 295-303.

Norman, F. and Martin, C. E. (1986). Effects of spine removal on Coryphantha vivipara in central Kansas. *Am. Midl. Nat.* 116: 118-124.

North, G. B., Lin Moore, T. and Nobel P. S. (1995). Cladode Development for *Opuntia ficus-indica* (Cactaceae) Under Current and Doubled CO_2 Concentrations. *AJB* 82: 159-166.

Nyffeler, R. 2002. Phylogenetic relationships in the cactus family (Cactaceae) based on evidence from trnK/matK and trnL-trnF sequences. *Am. J. Bot.* 89:312-326.

Ogawa, A., Eguchi, T. and Toyofuku, K. (2012). Cultivation methods for leafy vegetables and tomatos with low potassium content for dialysis patients. *Environ. Control Biol.* 50:407-414.

Osmond, C. B. (1978). Crassulacean acid metabolism: a curiosity in context. *Annu. Rev. Plant Physiol.* 29: 379-414.

Pimienta-Barrios, E., Zañudo-Hernandez, J., Rosas-Espinoza, V. C., Valenzuela-Tapia, A. and Nobel, P. S. (2005). Young daughter cladodes affect CO2 uptake by mother cladodes of *Opuntia ficus-indica*. *Ann. Bot.* 95:363-9.

Rebman, J. P. and Pinkava D. J. (2001). Opuntia cacti of North America— an overview. *Florida Entomologist* 84:474-483.

Sakamoto, Y., Watanabe, S., Nakashima, T. and Okano, K. (1999). Effects of salinity at two ripening stages on the fruit quality of single-truss tomato grown in hydroponics. *JHSB* 74:690-693.

Sánchez, D., Grego-Valencia, D., Terrazas, T. and Arias, S. (2015). How and why does the areole meristem move in Echinocereus (Cactaceae)? Ann. Bot. 115:19-26.

Santamaria, P. (2005). Nitrate in vegetables: toxicity, content, intake and EC regulation. *J. Sci. Food Agric.* 86:10-17.

Shetty, A. A., Rana, M. K. and Preetham, S. P. (2012). Cactus: a medicinal food. *J. Food Sci. Technol.* 49:530-6.

Shibutani, S, Kinoshita, K. (1968). Studies on the ecolological adaptation of lettuce. Faculty of Agriculture of Okayama University 32:25-34.

Silos-Espino, H., Valdez-Ortiz, A., Rascón-Cruz, Q., Rodríguez-Salazar, E. and Paredes-López, O. (2006). Genetic transformation of prickly-pear cactus (Opuntia ficus-indica) by *Agrobacterium tumefaciens. Plant Cell, Tissue and Organ Culture* 86:397-403.

Stefanelli, D., Winkler, S and Jones. R. (2011). Reduced nitrogen availability during growth improves quality in red oak lettuce leaves by minimizing nitrate content, and increasing antioxidant capacity and leaf mineral content. *Agricultural Sciences* 2:477-486.

Stintzing, F. C. and Carle, R. (2005). Cactus stems (Opuntia spp.): a review on their chemistry, technology, and uses. *Mol. Nutr. Food Res.* 49: 175-194.

Wahome, P. K., Oseni, T. O., Masarirambi, M. T. and Shongwe, V. D. (2011). Effects of different hydroponics systems and growing media on the vegetative growth, yield and cut flower quality of gypsophila (*gypsophila paniculata L.*) *World J. Agric. Sci.* 7: 692-698.

Wallace, R. S. 1995a. A family-wide phylogeny, subfamilial and tribal relationships, and suggestions for taxonomic realignments. *IOS Bulletin* 6:13.

Wallace, R. S. 1995b. Molecular systematic study of the Cactaceae: Using chloroplast DNA variation to elucidate cactus phylogeny. *Bradleya* 13:1-12.

Wallace, R. S. and Gibson, A. C. (2002). *Evolution and systematics.* Cacti: biology and uses. University of California Press, Berkeley, 1-21. In P. S. Nobel [ed.], *Cacti, biology and uses.* University of California Press, Los Angeles, California, USA.

Wang, H., W. Liang-Huan, W. Min-Yan, Z. Yuan-Hong, T. Qin-Nan and Z. Fu-Suo. (2007). Effects of Amino Acids Replacing Nitrate Growth, Nitrate Accumulation, and Macro element Concentrations in Pak-choi (Brassica chinensis L.). *Pedosphere* 17:595-600.

Yanagi, T., Yachi, T., Okuda, N. and Okamoto, K. (2006). Light quality of continuous illuminating at night to induce floral initiation of *Fragaria chiloensis L.* CHI-24-1. *Scientia Hort.* 109: 309-314.

Yanata, S. and Takata, K. (2014). Plant factory: the possible measures to revitalize the Wakayama's economy, regional studies series no. 43 (Revised Edition). Institute of Economic Research, Faculty of Economics, Wakayama University, 1-22.

Zhao, X., Yu, X., Foo, E., Symons, G. M., Lopez, J., Bendehakkalu, K. T., Xiang, J., Weller, J. L., Liu, X., Reid, J. B. and Lin, C. (2007). A study of gibberellin homeostasis and cryptochrome-mediated blue light inhibition of hypocotyl elongation. *Plant Physiol.* 145: 106-118.

In: Advances in Hydroponics Research ISBN: 978-1-53612-131-5
Editor: Devin J. Webster © 2017 Nova Science Publishers, Inc.

Expert Commentary

PLANT FACTORIES AND EDIBLE CACTI

*Takanori Horibe**
College of Bioscience and Biotechnology, Chubu University, Kasugai,
Aichi, Japan

ABSTRACT

Plant factories are horticulture greenhouses or automated facilities where vegetables and other crops can be produced throughout the year by controlling environmental conditions such as light, temperature, humidity, CO_2, and nutrient availability. Although food safety and constant supply of food crops to the market is the main advantages of plant factories, many new techniques have been recently invented to increase the productivity and nutritional content of vegetables and to produce high-quality and value-added products. Hydroponic culture is used as a basic cultivation system in plant factories. We have investigated the effects of hydroponic culture [deep flow technique (DFT)] and cultivation conditions on the growth of edible *Opuntia*, which is conventionally produced through soil or pot culture. Our study showed that the growth of cladodes was greatly affected by growth conditions in the plant factory. Edible *Opuntia* has many features suitable for the production in plant factories, which require large running costs. For

---

* Corresponding Author Email: t-horibe@isc.chubu.ac.jp. Fax: + 81-092-642-2913.

instance, it grows rapidly and can be vegetatively propagated through stems (in the short term until harvest). Additionally, *Opuntia* can be planted on cultivation panels at high density (effective use of available space) under relatively low light intensity using artificial light. Therefore, we propose that plant factories are powerful tools to cultivate and improve the quality of edible cacti.

INTRODUCTION

Plant factories can be defined as horticulture greenhouses or automated facilities where vegetables and crops can be produced throughout the year controlling environmental conditions, such as light, temperature, humidity, CO_2, and nutrient availability by ICT (Yanata et al., 2014; Hirama, 2015). There are two main types of plant factories. These include artificial lighting-based fully enclosed systems and natural sunlight-based systems. Hydroponic culture is mainly used as the cultivation system in enclosed plant factories, whereas both soil and hydroponic technologies are used in natural sunlight systems. Hydroponic culture can circumvent the expensive and time-consuming task of soil sterilization to prevent soil-borne diseases and enables precise fertilizer management (Wahome et al., 2011; Lakkireddy et al., 2012). Sunlight-based plant factories can use either exclusive natural sunlight or a combination of natural sunlight and artificial light. Such sunlight-based plant factories are widespread in Holland. Interest towards plant factories is increasing in East Asian countries such as China and Korea (Yanata et al., 2015).

Plant factories using artificial light have already been used in Japan for the production of leafy vegetables. The main advantages of modern plant factories are the food safety and steady supply of food crops to the market. Using artificial lighting-based fully enclosed plant factory, high quality pesticide-free vegetables can be produced throughout the year due to controlled, optimal cultivation environment (Yamori et al., 2014). Furthermore, because it does not need sunlight and natural soil, this type of plant factory can be built in any location and in any building. Hence, because its productivity is not dependent on climate and soil fertility, it enables food production even in regions where agriculture is not feasible. It is estimated that the number of plant factories in commercial production in Japan would exceed 120 by December 2012 (Kozai, 2013). One of the largest commercial plant factories in Japan can produce close to 25,000 leaf lettuce heads per day or 9 million heads per year (Kozai, 2013).

However, plant factories also have some disadvantages. For instance, initial investment and running cost (especially personnel, fuel, and light expenses) are very high (Yanata et al., 2014). Making profit through plant factories is not easy. It is roughly estimated that 20% of plant factories are making profit, 60% are breaking even, and 20% are operating at a loss (Kozai, 2013). Cultivating large number of plants as fast as possible and selling them at high price is required to ensure profit. Additionally, there are certain prerequisites for growing plants in plant factories for commercial production with artificial light. For the maximum use of the air space in enclosed type plant factories, plants raised there should be shorter than approximately30 cm in height because the distance between tiers is usually around 40 cm. In addition, plants should be able to grow well at relatively low light intensity and high planting density and with most of the plant parts (leaves, stems and roots) being edible or saleable at a high price. Therefore, most of the enclosed-type plant factories are used to grow leafy vegetables, herbs, root crops, medicinal plants, and other crops, which are short in height and can be sold at high prices; however, tomato, cucumber, eggplant, and seedlings of ornamental plants and bedding plants are produced in sunlight-based plant factories due to relatively low running cost compared to that in the enclosed-types (Kozai, 2013).

The development of modern plant factories is gaining momentum and advanced technologies are being adopted within plant factories for the cultivation of vegetables and crops. These include labor-saving features *via* automation (Park and Nakamura, 2015), the invention of automated environmental control systems (Oguntoyinbo et al., 2015), an improved light environment *via* artificial light (Li et al., 2016), and the "speaking plant" approach, which involves measuring the physiological responses of the plants using sensors for environmental optimization (Morimoto and Hashimoto, 2009). Some reports have shown that controlled environmental conditions in plant factories can also lead to improvement in the quality of vegetables and crops. For instance, Zhang (2015) reported that blue light-emitting diode (LED) irradiation increases the amount of ascorbic acid in citrus juice sacs. These new technologies lower the production cost and aid to cultivate high-value plants such as organic vegetables, functional vegetables, and medicinal plants.

Plant Factories May Be Powerful Tools for Producing High-Quality Edible Cacti

The stem of the cactus, *Opuntia* (genus *Opuntia*, subfamily *Opuntioideae*, family Cactaceae), is widely consumed as a vegetable in Mexico and Mediterranean countries (Stintzing and Carle, 2005; Cruz–Hernández and Paredes–López, 2010). These plants are also used in some countries as remedies for a variety of health problems (El-Mostafa et al., 2014). For example, prickly pear is used as a folk medicine for burns, edema, and indigestion (Shetty et al., 2012). In Japan, edible *Opuntia* plants are also produced as vegetables, mainly in Kasugai city, Aichi Prefecture.

Opuntia is characterized by remarkable adaptation to arid and semi-arid climates. It exhibits crassulacean acid metabolism (CAM), a CO_2 concentrating mechanism that potentially leads to higher optimal temperatures for photosynthesis (Monson 1989). CAM plants also have certain anatomical modifications that enable them to survive droughts, such as thick cuticles and low stomatal frequency, together with night-time CO_2 uptake (Drennan and Nobel 2000; Pimienta-Barrios et al., 2005). With respect to its growth behavior, daughter cladodes develop from the areole of the mother cladode and this process is repeated (Pimienta-Barrios et al., 2005). Daughter cladodes show C3 photosynthesis with daytime stomatal opening during the early stages of development (Osmond 1978; Acevedo et al., 1983) and import water from the mother cladodes (Pimienta-Barrios et al., 2005). The acidity in cladodes fluctuate even after harvesting, affecting its taste (Cantwell et al., 1992).

Edible cacti are commonly grown in soil or pot culture. Major problems encountered in growing vegetables in soil include soil-borne diseases, salt accumulation, and difficulty in fertilizer management (Lakkireddy et al., 2012). Hydroponics is a method of growing plants using nutrient solution (water and fertilizer) with or without the use of an artificial medium. It can be used for the production of edible *Opuntia* with added advantages. For example, hydroponics can be used to reduce the nitrate content of vegetables (Wang et al., 2007; Stefanelli et al., 2011), which can be converted to harmful nitrites in human, and potassium content (Ogawa, 2012), amount of which is restricted for dialysis patients. It is also effective in improving quality of crops as well as precisely controlling the fertilizers (Sakamoto et al., 1999).

We have shown that edible *Opuntia* can be cultivated by DFT hydroponic system and hydroponics may be effective means to decrease the number of spines on cladodes (Horibe and Yamada, 2016). Although a number of beneficial functions have been attributed to the spines on cacti, including mechanical protection from herbivores (Norman and Martin 1986), shading of the stem (Nobel et al., 1986), and collection of fog to absorb water (Ju et al., 2012), spines on edible cacti are one of the most undesirable characteristics for consumers. They are usually burnt or removed using a knife before the cactus is put on sale. Therefore, techniques and tools, which lead to spine elimination, can improve commercial value of edible cacti. In this study, we have cultivated edible *Opuntia* in enclosed-type plant factory located in our university (Figure 1) and studied the effects of red and blue LED light on the growth and spine occurrence in daughter cladodes. We have shown that light wavelength strongly affects the growth and number of spines in daughter cladodes (Horibe et al., 2016). Thus, it seems that the growth and development of spine of edible cacti can be manipulated by regulating the environmental conditions, which in turn will be effective for improving the growth and quality of edible cactus.

Overall, plant factories have many advantages for improving the quality and steady supply of vegetables and crops. In addition, edible cacti have many suitable features for production in plant factories. For instance, they grow rapidly and can be vegetatively propagated using stems (in the short term until harvest). Additionally, they can be planted on cultivation panels at high density (effective use of available space) and cultivated under relatively low light intensity using artificial light (Horibe et al., 2016). The cactus plants are consumed in many countries and have various health benefits. Currently, people have strong interest in the safety and health benefits of foods in Japan. Hence, there may be a greater demand in Japan as well as other countries for high quality, pesticide-free, and functional edible cacti produced in plant factories. Therefore, stable production of edible *Opuntia* with increased nutritive value through environment control and hydroponics may be practiced in the near future.

Figure 1. Hydroponic system in fully enclosed-type plant factory.

CONCLUSION

Plant factories have strong advantages in producing high quality vegetables throughout the year through controlled environmental conditions. Edible cacti are consumed as healthy vegetable in many countries and exhibit various features suitable for production in plant factories. We have shown that edible cactus can be efficiently cultivated by hydroponic culture with controlled environmental conditions, which can lead to improvement in its quality. Consequently, we propose that plant factories are powerful tools to cultivate edible cacti and improve their quality, although more cumulative evidence is necessary.

REFERENCES

Cantwell, M., Rodríguez-Felix, A. and Robles-Contreras, F. (1992). Postharvest physiology of prickly pear cactus stems. *Scientia Horticulturae* 50:1-9.

Cruz-Hernández, A. and Paredes-López, O. (2010). Enhancement of economic value of nopal and its fruits through biotechnology. *JPACD* 12: 110-126.

Drennan P. M. and Nobel P. S. (2000). Responses of CAM species to increasing atmospheric CO_2 concentration. *Plant, Cell and Environ.* 23:767-781.

El-Mostafa, K., El-Kharrassi, Y., Badreddine, A., Andreoletti, P., Vamecq, J., El-Kebbaj, M. S., Latruffe, N., Lizard, G., Nasser, B. and Cherkaoui-Malki, M. (2014). Nopal cactus (*Opuntia* ficus-indica) as a source of bioactive compounds for nutrition, health and disease. *Molecules* 19:14879-14901.

Hirama, J. (2015). The history and advanced technology of plant factories. *Environ. Control Biol.* 53:47-48.

Horibe, T. and Yamada, K. (2016). Hydroponics culture of edible *Opuntia* 'Maya': drought stress affects the development of spines on daughter cladodes. *Environ. Control Biol.* 54: 153-156.

Horibe, T., Iwagaya, Y., Kondo, H. and Yamada, K. (2016). Hydroponics Culture of Edible *Opuntia* 'Maya': Effect of Constant Red and Blue Lights on Daughter Cladodes Growth and Spine Development. *Environ. Control Biol.* 54: 165-169.

Ju, J., Bai, H., Zheng, Y., Zhao, T., Fang, R. and Jiang, L. A. (2012). Multi-structural and multi-functional integrated fog collection system in cactus. *Nat. Commun.* 3: 1247.

Kozai, T. (2013). Plant factory in Japan - current situation and perspectives. *Chronica Horticulturae* 53: 8-11.

Lakkireddy, K. K. R., Kasturi K. and Sambasiva, R. K. R. S. (2012). Role of hydroponics and aeroponics in soilless culture in commercial food production. *JAST* 1: 26-35.

Li, K., Li, Z. and Yang, Q. (2016). Improving light distribution by zoom lens for electricity savings in a plant factory with light-emitting diodes. Front. Plant Sci. 7:92. doi: 10.3389/fpls.2016.00092.

Monson R. K. (1989). On the evolutionary pathways resulting in C_4 photosynthesis and Crassulalacean acid metabolism (CAM). *Adv. in Ecol. Res.* 19:57-110.

Morimoto, T. and Hashimoto, Y. (2009). Speaking plant/fruit approach for greenhouses and plant factories. *Environ. Control Biol.* 47:55-72.

Nobel, P. S., Geller, G. N., Kee, S. C. and Zimmerman, A. D. (1986). Temperatures and thermal tolerances for cacti exposed to high-temperatures near the soil surface. *Plant Cell Environ.* 9: 279-287.

Norman, F. and Martin, C. E. (1986). Effects of spine removal on Coryphantha vivipara in central Kansas. *Am. Midl. Nat.* 116: 118-124.

Ogawa, A., Eguchi, T. and Toyofuku, K. (2012). Cultivation methods for leafy vegetables and tomatoes with low potassium content for dialysis patients. *Environ. Control Biol.* 50:407-414.

Oguntoyinbo, B., Saka, M., Umemura, Y. and Hirama, J. (2014). Plant factory system construction: cultivation environment profile optimization. *Environ. Control Biol.* 53:77-83.

Park, J. E. and Nakamura, K. (2015). Automatization, labor-saving and employment in a plant factory. *Environ. Control Biol.* 53: 89-92.

Pimienta-Barrios, E., Zañudo-Hernandez, J., Rosas-Espinoza, V. C., Valenzuela-Tapia, A. and Nobel, P. S. (2005). Young daughter cladodes affect CO2 uptake by mother cladodes of *Opuntia ficus-indica*. *Ann. Bot.* 95:363-9.

Sakamoto, Y., Watanabe, S., Nakashima, T. and Okano, K. (1999). Effects of salinity at two ripening stages on the fruit quality of single-truss tomato grown in hydroponics. *JHSB* 74:690-693.

Shetty, A. A., Rana, M. K. and Preetham, S. P. (2012). Cactus: a medicinal food. *J. Food Sci. Technol.* 49:530-6.

Stefanelli, D., Winkler, S and Jones. R. (2011). Reduced nitrogen availability during growth improves quality in red oak lettuce leaves by minimizing nitrate content, and increasing antioxidant capacity and leaf mineral content. *Agricultural Sciences* 2:477-486.

Stintzing, F. C. and Carle, R. (2005). Cactus stems (*Opuntia* spp.): a review on their chemistry, technology, and uses. *Mol. Nutr. Food Res.* 49: 175-194.

Wahome, P. K., Oseni, T. O., Masarirambi, M. T. and Shongwe, V. D. (2011). Effects of different hydroponics systems and growing media on the vegetative growth, yield and cut flower quality of gypsophila (*gypsophila paniculata L.*) *World J. Agric. Sci.* 7: 692-698.

Wang, H., W. Liang-Huan, W. Min-Yan, Z. Yuan-Hong, T. Qin-Nan and Z. Fu-Suo. (2007). Effects of Amino Acids Replacing Nitrate Growth, Nitrate Accumulation, and Macro element Concentrations in Pak-choi (Brassica chinensis L.). *Pedosphere* 17:595-600.

Yamori, W., Zhang, G., Takagaki, M. and Maruo, T. (2014). Feasibility study of rice growth in plant factories. *J. Rice Res.* 2. doi: 10.4172/jrr.1000119.

Yanata, S., Nomakuchi, T. and Ishibashi, K. (2015). Plant factory engineering strategy of Japanese manufacturer and agri-business. *Journal of Agricultural Chemistry and Environment* 4: 15-19.

Yanata, S. and Takata, K. (2014). *Plant factory: the possible measures to revitalize the Wakayama's economy*, regional studies series no. 43 (Revised Edition). Institute of Economic Research, Faculty of Economics, Wakayama University, 1-22.

Zhang, L., Ma, G., Yamawaki, K., Ikoma, Y., Matsumoto, H., Yoshioka, T., Ohta, S. and Kato, M. (2015). Regulation of ascorbic acid metabolism by blue LED light irradiation in citrus juice sacs. *Plant Sci.* 134-42.

INDEX

A

acclimatization, x, 91, 92, 130
acid, x, 42, 44, 51, 53, 61, 74, 80, 82, 83, 86, 111, 112, 115, 118, 123, 155, 167, 169, 170, 176, 180
acidity, 176
active oxygen, 66
adaptation, 12, 154, 155, 171, 176
Advantages, 75, 106, 107, 129, 131
Aeroponics, 103, 105
agriculture, xi, 24, 52, 102, 103, 106, 107, 123, 124, 125, 126, 128, 133, 138, 148, 175
Agrobacterium, 171
air temperature, 11, 18, 39, 40
alcohol production, 93, 97
alfalfa, 129
algae, 95
aliphatic compounds, 76
alternative fertilizer solution, 92
alternative nutritional solution, 138, 141
amino, 78, 86, 111, 112, 118
amino acid, 78, 111, 112, 118
Amino acid compositions, 115
amino acids, 78, 111, 112, 118
amino groups, 86

ammonia, 36, 39, 112, 144
Ammoniacal nutrition, 34
ammonium, vii, ix, 33, 34, 35, 36, 38, 39, 40, 45, 46, 48, 50, 51, 52, 53, 54, 55
Ammonium Phytotoxicity, 45
anchovy wastewater, 112
ANOVA, 61
antibacterial, 116
antifungal activities, 116
antioxidant, vii, ix, 51, 52, 53, 58, 59, 61, 63, 64, 66, 68, 69, 70, 71, 81, 110, 111, 116, 124, 127, 171, 181
antioxidant enzymes, 51, 52, 58, 59, 63, 64, 68
Application, vi, 28, 29, 98, 110, 153
application techniques, 4
aquaculture, 110
Aquaponics, 106
Arabidopsis thaliana, 34, 88, 123
arsenic, 74, 83, 86, 109
ascorbic acid, 176, 182
assessment, 21, 98
assimilation, 34, 47, 50, 51
atmosphere, 19, 77

Index

B

Bacillus species, 113
bacteria, 76, 78, 130, 133
barley, 43, 50, 53, 112, 117, 118
bedding, 103, 175
beneficial effect, 35, 111
benefits, 42, 43, 139, 167, 178
bioaccumulation, 59, 69, 87, 88
bioactive compounds, 111, 119, 121, 168, 179
bioavailability, 109
biochemical processes, 52
Bio-concentration factor, 74
bioconversion, 124, 128
biodegradation, 76, 112, 120
biodiversity, 98, 102, 103, 124
biofertilizer, xi, 102, 112, 116, 119, 120, 121, 135, 145
biogas, 94, 95
biological sciences, 54
biomarkers, x, 58, 67
biomass, x, 74, 75, 82, 83, 84, 85, 86, 122, 156
bioremediation, 76, 94
biosynthesis, 78
biotechnology, 125, 167, 168, 179
Brazil, 33, 35, 91, 93, 137, 138, 141
breakdown, 75, 76
building-type hydroponics, xi, 101, 102, 120
By-product, 97
by-products, xi, 92, 93, 94, 97, 149, 150

C

Cacti, vi, 153, 154, 156, 171, 173, 176
cadmium, 52, 69, 70, 74, 110
calcium, 36, 44, 140, 146
calibration, 4, 23, 30, 61
CAM, 155, 167, 168, 169, 176, 179, 180
capillary, 105
Capillary Action Technique, 105
carbohydrates, 113, 128, 129, 146
carbon, 50, 53, 55, 112, 122
carbon dioxide, 112
carbonaceous oxygen demand, 112
carboxyl, 86
carboxylic acid, 133
cassava wastewater, vi, vii, xi, xii, 92, 95, 97, 137, 138, 140, 141, 142, 143, 144, 145, 146, 147, 148, 149, 150
catalase, vii, ix, 49, 58, 61, 81
cation, 34, 36, 40, 109
centella asiatica, 58
challenges, 106, 134, 139
chemical, xi, xii, 41, 75, 92, 94, 97, 112, 116, 133, 138, 140, 142, 143, 144
chemicals, 2, 75, 103, 107
children, ix, 33, 36
China, 32, 71, 128, 174
chlorophyll, 45, 48, 50, 69, 110, 116, 146
chloroplast, 34, 48, 171
choline, 44, 53
chromatograms, x, 74, 83
chromatographic and instrumental analysis, 83
chromium, x, 50, 73, 74, 79, 80, 81, 82, 83, 84, 85, 86, 88
Chromium(VI), 74, 82, 83
climate, viii, ix, 1, 2, 6, 13, 17, 24, 34, 102, 106, 110, 128, 174
climate change, 34, 102, 106
climates, 92, 154, 176
climatic factors, viii, 2, 19
CO_2, xiii, 34, 51, 55, 155, 157, 167, 170, 173, 174, 176, 179, 181
commercial, xi, 17, 36, 42, 92, 96, 97, 107, 109, 111, 112, 114, 115, 116, 117, 118, 124, 126, 156, 159, 160, 166, 169, 175, 177, 180
competition, ix, 34, 58, 65
complement, 139

Index

composition, 29, 51, 53, 109, 111, 112, 118, 139, 140
composting, 106, 134
compounds, 45, 111, 115, 119, 120, 129, 145, 155, 168, 179
conductance, 45, 49, 55
conductivity, 9, 10, 11, 23, 107, 143
constituents, 75, 130
construction, 103, 180
consumers, 75, 157, 163, 167, 177
consumption, 3, 4, 6, 36, 94, 110, 119, 121, 139, 155
containers, 9, 105, 166
contaminant, 75, 76, 77
contaminated soil, 59, 69, 71, 87, 88
contaminated water, 77
contamination, x, 73, 107
cooling, 3, 6, 17, 31
coordination, 80, 85
copper, 69, 70, 71
Corn steep liquor, 96
cost, x, 74, 75, 93, 97, 105, 106, 126, 164, 166, 175, 176
cotton, 43, 45, 52, 147
crop, viii, ix, x, 1, 2, 3, 4, 5, 6, 7, 8, 10, 11, 12, 13, 15, 16, 17, 18, 20, 22, 23, 26, 27, 28, 30, 31, 33, 36, 37, 41, 59, 70, 92, 93, 94, 96, 103, 107, 108, 113, 126, 128, 130, 132, 144, 148, 154, 166, 168, 169, 173, 174, 175, 177
crop production, x, 92, 107, 128, 130
Cu, vii, ix, 58, 59, 60, 61, 62, 63, 64, 65, 66, 67, 68, 69, 71, 78, 96, 134, 146
cultivars, 28, 155, 156
cultivation, vii, xii, xiii, 2, 3, 9, 12, 13, 15, 16, 18, 19, 22, 23, 24, 36, 39, 42, 52, 92, 93, 94, 97, 103, 108, 120, 138, 140, 142, 146, 149, 153, 156, 157, 159, 160, 161, 163, 166, 167, 173, 174, 175, 178, 180
cultivation conditions, xiii, 160, 163, 173
culture, vii, viii, xi, xii, xiii, 1, 2, 12, 18, 27, 28, 29, 54, 69, 86, 87, 92, 94, 95, 96, 98, 99, 102, 106, 107, 108, 109, 111, 112, 114, 115, 116, 117, 118, 119, 120, 121, 124, 126, 127, 129, 131, 132, 138, 145, 153, 156, 157, 158, 159, 160, 163, 164, 165, 166, 168, 169, 173, 174, 177, 179, 180
culture conditions, 106, 112
culture medium, 94, 98, 99
cysteine, 78, 111
cysteine-rich protein, 78
cytoplasm, 55, 65

D

daily intake limit, 36
decomposition, 150
Deep Flow Technique, 104
defence, 59, 66, 70, 125
defense mechanisms, 66
deficiency, 34, 50, 51, 55, 65, 93, 107, 146
deficit, viii, 2, 6, 13, 15, 20, 22, 30, 43
degradation, 35, 48, 74, 76, 107, 125, 126
detection, 4, 17, 21, 26, 27
detoxification, 78, 81, 89
DFT, xiii, 157, 158, 159, 162, 173, 177
dialysis, 163, 170, 177, 180
direct measure, viii, 2, 26
Disadvantages, 106, 107, 108
diseases, 106, 107, 113, 133, 156, 174, 177
disinfection, 125
dissolved oxygen, 94
distribution, 120, 168, 180
DNA, 169, 171
drainage, viii, 1, 3, 5, 7, 22, 27, 102, 110
drinking water, 139
drought, 51, 128, 159, 168, 179
dry matter, 34, 35, 46, 47, 55
drying, 104, 120

E

E. coli, 125
E. crassipes (Eichhornia crassipes), 74, 79, 80, 81, 82, 83, 84, 85, 86, 87
East Asia, 67, 174
eco sanitation, 138, 139
eco-friendly, 106, 116, 120, 121, 139
edema, 155, 176
Edible Opuntia, xiii, 155, 168, 173, 180
effluent, xi, 92, 94, 97, 143, 147
effluents, vii, xii, 138, 139
electric current, 146
electrical conductivity, xii, 8, 9, 10, 23, 31, 105, 107, 138, 142, 143, 144, 145, 146, 160, 165
electromagnetic waves, 20
elongation, 113, 159, 160, 161, 165, 169, 172
emission, xi, 5, 27, 102, 121, 160
employment, 140, 180
energy, viii, 2, 3, 7, 13, 31, 34, 36, 39, 50, 76, 95, 107, 131, 146
energy consumption, 34
energy efficiency, 131
energy expenditure, 34
energy input, 107
energy transfer, 146
engineering, 78, 181
environment, viii, xii, 1, 2, 5, 7, 13, 16, 17, 23, 24, 58, 75, 76, 88, 98, 102, 103, 106, 107, 108, 109, 130, 133, 138, 139, 147, 150, 154, 157, 160, 162, 166, 174, 175, 178, 180
environment control, 157, 178
environmental conditions, viii, xiii, 2, 15, 34, 78, 154, 157, 160, 165, 166, 173, 174, 176, 177, 179
environmental effects, 139
environmental factors, 160

environmental impact, xii, 128, 138, 139, 147
Environmental Protection Agency (EPA), 134
environmental stress, 70, 113
Environmentally friendly, 75
environments, 15, 154
enzymatic activity, 134
enzymes, ix, 35, 47, 48, 50,51, 52, 58, 59, 61, 63, 64, 65, 66, 67, 68, 69, 70, 76, 78, 121, 146
equipment, viii, 2, 13, 20, 105, 127
erosion, 77, 102
ethanol, 93, 97
Europe, 103, 123
European Union, 36, 37
evaporation, 3, 7, 11, 12, 13, 120, 121
evapotranspiration, viii, 2, 3, 5, 11, 12, 13, 15, 16, 24, 25, 26, 27, 29, 30
evidence, 65, 170, 179
exposure, vii, ix, 58, 62, 64, 65, 66, 69, 79, 80, 86
extraction, x, 61, 73, 74, 79, 82, 140, 146
extracts, 68, 113, 116, 123, 125, 126, 146

F

factories, viii, xii, 58, 157, 166, 168, 173, 174, 175, 177, 179, 180, 181
farmers, xii, 2, 153
farms, 125, 131, 147
Feasibility Test, vi, 101
feces, 106, 139, 144
fermentation, 141
ferredoxin, 34
fertilization, 36, 127, 140, 144, 148
fertilizer sources, 115
fertilizers, xi, 36, 37, 58, 67, 97, 109, 111, 112, 113, 114, 115, 116, 118, 121, 124, 130, 138, 177
filtration, xi, 92, 95

financial, 67, 108
financial support, 67
fish, 103, 106, 111, 112, 114, 115, 117, 119, 120, 121, 122, 123, 124, 125
Fish protein hydrolysates, 111
Fish Waste, 111
fishery waste, vii, xi, 101, 102, 110, 111, 120, 121, 128
fixation, 110
flavonoids, 130
Floating Technique, 105
flora and fauna, 95
flour, 94, 138, 140, 141, 147
flour houses, 138, 147
flowers, 103, 105, 155
fluctuations, 106
fluorescence, 22, 50
food, xiii, 54, 59, 75, 102, 103, 106, 110, 123, 129, 132, 168, 169, 171, 173, 174, 180, 181
food chain, 59, 75
food production, 102, 168, 169, 175, 180
food safety, xiii, 106, 173, 174
food security, 103, 132
formation, 78, 81, 130, 146
free radicals, 81
freshwater, 103
fruits, 10, 113, 119, 120, 121, 167, 179
fruits and vegetables, 119, 120, 121
FTIR, x, 73, 85, 86
FTIR technique, x, 73
functional food, 129
fungi, 76, 78

G

gene expression, 130
genes, 88, 122, 146
genus, 154, 176
germination, 113, 116, 126, 127
gibberellin, 162, 172

glucosinolates, 130
Greece, 1, 14, 30, 131
green alga, 75, 78
green color index, 48
greenhouse, viii, 1, 2, 3, 7, 11, 12, 13, 15, 16, 17, 19, 20, 22, 23, 24, 25, 26, 27, 28, 29, 30, 31, 39, 40, 54, 60, 126, 127, 128, 131, 132, 133, 141, 151, 162, 165, 169, 173, 174, 180
groundwater, 3, 75, 99, 107
growth, viii, x, xii, xiii, 1, 2, 4, 5, 6, 8, 9, 11, 12, 15, 16, 19, 23, 26, 27, 28, 29, 34, 38, 49, 50, 51, 52, 53, 55, 59, 65, 69, 70, 73, 75, 91, 99, 102, 107, 108, 109, 110, 111, 112, 116, 118, 122, 123, 125, 126, 128, 130, 132, 133, 146, 153, 155, 156, 157, 159, 160, 162, 165, 166, 167, 171, 173, 176, 177, 181
growth rate, 9, 23, 111, 159, 165
growth temperature, 163
guidelines, 123

H

H_4SiO_4, 42
habitat, 102
harmful effects, 2, 59
harvesting, 8, 12, 102, 166, 176
hazardous waste, 74
health, 104, 107, 112, 155, 163, 167, 168, 176, 178, 179
health problems, 155, 163, 176
health risks, 104
heavy metals, vii, x, 59, 69, 73, 74, 79, 88, 93, 122
height, 16, 47, 48, 164, 175
heterosis, 70
hexavalent chromium (Cr), 75
high temperatures, 28, 36, 38
high-value crops, 108
history, 127, 128, 168, 179

homeostasis, 68, 172
homogeneity, 13
horticultural crops, 26, 29
human, 59, 102, 104, 139, 155, 163, 168, 177
human health, 104
human urine, vi, vii, xii, 137, 138, 140, 141, 142, 143, 144, 145, 146, 147, 148, 150, 151
humidity, xiii, 3, 11, 157, 160, 163, 173, 174
Hydrochromate, 80
hydrogenation, 144
hydroponics, vii, ix, x, xi, xii, 30, 33, 36, 41, 42, 47, 91, 92, 93, 95, 96, 98, 101, 102, 103, 106, 107, 109, 110, 112, 116, 119, 120, 121, 123, 124, 125, 127, 130, 131, 132, 153, 157, 159, 160, 163, 169, 170, 171, 177, 178, 180, 181
hydroponics culture, 157, 160
Hydroponics Culture, 168, 180
Hydroponics Model, 119
hydroxide, 83
hydroxyl, 86
Hyperaccumulation, 76
hypocotyl, 162, 172

I

ideal, xii, 111, 153, 166
improvements, 44
in vitro, 53, 67, 130, 131
industrial residue, 92
industrialization, 149, 150
industry, 93, 94, 97, 98, 122, 167
infection, 19, 108, 121
Infrared spectra, 85
inhibition, 15, 61, 65, 121, 172
inoculation, 110, 123
integrators, 23
integrity, 145

interface, 68
investment, 107, 108, 175
investments, 16
ions, x, 34, 38, 43, 74, 75, 79, 80, 81, 82, 83, 84, 85, 86
irradiation, 39, 113, 127, 160, 161, 162, 176, 182
irrigation, vii, viii, ix, 1, 2, 3, 4, 5, 6, 7, 8, 9, 10, 11, 13, 16, 17, 18, 20, 22, 23, 24, 25, 26, 27, 28, 29, 30, 31, 103, 108, 133

J

Japan, 54, 84, 153, 155, 156, 157, 160, 165, 173, 174, 176, 178, 180
Jordan, 98

K

K^+, 109, 122
kidney, 117, 118
kill, x, 92
kinetic model, 87
kinetics, 87, 109
Korea, 101, 122, 174

L

Latin America, 149, 151, 155
leaching, 23, 31, 75, 133
lead, ix, 10, 23, 33, 34, 45, 59, 68, 69, 70, 74, 111, 156, 166, 176, 177, 179
leakage, 75
LED, 120, 135, 160, 176, 177, 182
life cycle, 109
light, xiii, 6, 15, 22, 39, 48, 51, 52, 60, 105, 140, 157, 160, 161, 162, 165, 169, 172, 173, 174, 175, 177, 178, 180, 182
light conditions, 6, 162
Light Environment, 160

Light Irradiation, 39
light-emitting diodes, 180
lipids, 113
liquid chromatography, 79, 89
low temperatures, 155
LSD, 86
Luo, 66, 69

M

mackerel wastewater, 112, 128
macronutrients, 145, 146
magnesium, 44, 140, 146
majority, 3, 16, 17
Malaysia, 57, 60, 67
management, vii, viii, xii, 1, 3, 10, 13, 27, 30, 106, 107, 108, 121, 125, 131, 132, 133, 151, 153, 156, 163, 166, 174, 177
manganese, 53, 110
manipulation, 108
mass, 4, 11, 16, 41, 47, 48, 79, 86, 89, 140, 146, 166
mass spectrometry, 79, 89
materials, x, 29, 74, 83, 84, 85, 86, 96, 111, 166
measurements, 3, 8, 11, 12, 16, 18, 19, 20, 23, 26, 27, 30, 109
media, 16, 30, 31, 106, 111, 171, 181
medicine, 59, 155, 176
Mediterranean climate, 25
Mediterranean countries, 131, 155, 176
melon, 12, 43
membrane separation processes, 122
meristem, 155, 170
metabolism, 47, 50, 53, 55, 69, 76, 112, 131, 146, 155, 167, 169, 170, 176, 180, 182
metabolites, 118, 121, 125, 127
metal ion, x, 73, 78
metals, ix, 58, 59, 60, 61, 62, 64, 65, 66, 74, 75, 78, 80, 139

methemoglobinemia, 36
methodology, xii, 138
Mexico, 125, 128, 134, 154, 176
microclimate, 4
micronutrients, 34, 44, 70, 140, 146
microorganisms, 74, 75, 76, 78, 97, 144
mildew, 122
mitigator, 43, 47
mixed-type, 120, 121
modelling, 26, 27, 30
models, viii, 2, 3, 10, 11, 13, 15, 16, 20, 23, 25
modifications, 155, 176
MODIS, 32
moisture, 8, 9, 10, 25, 28, 93
molecular weight, 78
molecules, 44, 92, 116
morphogenesis, 160
morphology, 84, 109, 154
multiple factors, 11

N

NaCl, 29, 54, 131
necrosis, 166
negative effects, 45
Netherlands, 51, 52, 70
NH_4^+, xii, 34, 35, 36, 37, 38, 39, 40, 41, 43, 45, 46, 47, 49, 50, 51, 52, 55, 66, 68, 138, 143
nitrate, ix, 27, 33, 34, 36, 37, 47, 51, 53, 54, 55, 112, 133, 163, 171, 177, 181
Nitrate-Ammonium Ratios, v, 33, 36, 38, 41
nitrates, 3, 54
nitric oxide, 66
nitrifying bacteria, 106
nitrite, 34, 54
nitrogen, vii, ix, xii, 33, 34, 36, 37, 39, 45, 46, 47, 50, 51, 52, 53, 55, 70, 86, 95, 110, 112, 113, 123, 124, 130, 131, 138, 140, 143, 144, 145, 146, 163, 171, 181

nitrogen fixation, 113
nitrogenous oxygen demand, 112
N-N, xii, 37, 138, 143, 144, 145
NO$_3^-$, 34, 35, 36, 37, 38, 39, 41, 42, 47, 52, 55
NO$_3^-$/NH$_4^+$, 34, 35, 38, 41, 42
non-hazardous, 116
nutrient, vii, viii, ix, x, xii, xiii, 1, 2, 5, 7, 10, 11, 22, 24, 29, 30, 33, 34, 35, 36, 37, 38, 39, 40, 41, 42, 43, 44, 45, 48, 49, 51, 55, 60, 66, 70, 82, 88, 91, 92, 93, 94, 95, 96, 97, 98, 103, 104, 105, 106, 107, 108, 109, 110, 116, 119, 120, 121, 124, 126,133, 138, 139, 140, 142, 143, 146, 147, 148, 153, 156, 157, 160, 165, 173, 174, 177
Nutrient Film Technique, 104
nutrients, x, xi, 30, 40, 41, 44, 49, 60, 70, 91, 92, 93, 94, 95, 96, 102, 103, 104, 105, 110, 116, 119, 138, 139, 140, 145, 146, 147
nutrition, 28, 34, 35, 36, 38, 39, 41, 54, 55, 68, 76, 109, 111, 112, 119, 128, 129, 130, 168, 179
nutritional imbalance, 45
nutritive solution, vi, vii, xi, 92, 93, 95, 96, 98, 132, 137, 142, 145, 149

O

Oceania, 27
OECD, 88
oil, 92, 132
oil production, 132
Oligomers, 113
oligosaccharide, 127
opportunities, 132
optimization, 175, 180
organic compounds, 34
organic matter, 94
ornamental plants, 10, 175

orthosilicic acid, 42, 53
OSA, 42, 44, 45
osmosis, 124
osmotic pressure, 108
Osmotic Regulation, 49
oxidation, 61, 80, 112
oxidative stress, 45, 48, 49, 52, 66, 71
oxygen, xii, 69, 70, 79, 94, 104, 105, 112, 138, 142, 143, 144, 166

P

pathogens, 35, 97, 104, 108, 115, 121, 134, 139
pathways, 162, 169, 180
Pb (Lead), v, vii, ix, 57, 58, 59, 60, 61, 62, 63, 64, 65, 66, 67, 69, 70, 71, 74, 78
peat, 12, 14, 23, 103
peptides, 87, 129
permeability, 145
permeable membrane, 95
pesticide, 106, 174, 178
pests, 106, 107, 113, 116, 133
pH, xii, 34, 39, 44, 45, 51, 54, 60, 80, 95, 96, 105, 107, 138, 141, 142, 143, 144, 145, 151
pH (potential of hydrogen), xii, 34, 39, 44, 45, 51, 54, 60, 80, 95, 96, 105, 107, 138, 141, 142, 143, 144, 145, 151
phenolic compounds, 95, 111
phosphate, 37, 79, 80
phospholipids, 146
phosphorus, xii, 70, 95, 112, 113, 130, 131, 138, 140, 142, 143, 144, 145, 146
phosphorylation, 34, 53
photosynthesis, 4, 10, 22, 47, 50, 52, 53, 88, 146, 155, 161, 168, 169, 176, 180
Photosynthetic Activity, 48
photosynthetic performance, 28
physiology, 109, 179
Phytoaccumulation, v, 73, 74

Index

phytochelatin, 78
phytoremediation, 74, 75, 78, 79, 86, 87, 88, 126
Phytoremediation technique, 79
Phytostabilization, 77
Phytovolatization, 77
plant growth, 5, 10, 30, 38, 49, 69, 92, 102, 103, 109, 111, 112, 113, 118, 119, 120, 123, 130, 146, 160, 163
plant monitoring, 17, 18, 20, 24
Plant sensing, 17
plants, ix, x, xi, xii, 4, 6, 11, 12, 13, 14, 16, 17, 18, 22, 23, 24, 30, 31, 33, 34, 35, 38, 39, 41, 42, 43, 44, 45, 46, 47, 49, 50, 51, 52, 53, 54, 55, 58, 59, 60, 61, 65, 66, 67, 68, 69, 70, 71, 74, 75, 76, 77, 78, 79, 81, 87, 88, 89, 91, 92, 93, 95, 103, 104, 105, 106, 107, 108, 109, 110, 111, 113, 114, 116, 118, 119, 121, 122, 123, 125, 126, 127, 129, 132, 134, 135, 138, 140, 146, 153, 154, 155, 156, 162, 163, 166, 167, 168, 175, 176, 177, 178
pollutants, x, 73, 93, 104
pollution, xi, 71, 74, 93, 94, 96, 110, 125, 138
polymerization, 44
polymers, 44
polyphenols, 130
polysaccharide, 86, 113, 134
polyunsaturated fat, 81
polyunsaturated fatty acids, 81
population, 75, 102, 111
Portugal, 148
Possible Problems, 120
potassium, xii, 34, 36, 40, 43, 44, 50, 53, 113, 127, 130, 138, 140, 143, 144, 145, 146, 163, 170, 177, 180
potassium silicate, 43, 44, 53
potato, 94, 131
Practical Use, 109
precipitation, 44, 45, 77, 154
preparation, 36, 37, 44, 82, 103, 107, 141

production costs, xi, 92
prokaryotes, 78
propagation, x, 92, 164
propylene, 44
protection, 2, 115, 122, 123, 134, 156, 159, 177
protein hydrolysates, 111, 123
proteins, 34, 43, 113, 122, 129, 146, 169
pruning, 12, 24
PVC, 95, 104

R

radiation, viii, 2, 3, 6, 7, 8, 10, 11, 13, 15, 16, 18, 20, 27, 28, 29, 30, 52, 129
reactions, 53, 81, 102, 131
reactive oxygen, 35, 59, 66, 81
Reactive Oxygen Species, 48
reading, 63, 64
real time, viii, 2, 3, 6, 11, 13, 16, 17, 24
recycling, xi, xii, 102, 138, 140, 147
red bean, 112
reflectance indices, 4, 21, 24, 27
regeneration, 168
regulatory agencies, ix, 33
remediation, 74, 75
remote sensing, 20, 30
reproduction, 167
requirements, 16, 21, 109, 146
researchers, 4, 17, 147
resistance, 27, 31, 70, 78, 109, 113, 115, 116, 124, 134
resources, 110, 121
respiration, 146
response, 18, 19, 21, 28, 41, 55, 69, 115, 124, 130, 134, 162
restrictions, 17
Reuse, 110, 148, 149
reverse osmosis, 104
Rhizodegradation, 76
Rhizofiltration, 77

Rhizopus, 134
Rhodophyta, 113
rhythmicity, 167
rice husk, 104
risk, 34, 36, 59, 108, 144
root, 5, 8, 9, 10, 15, 22, 23, 49, 55, 61, 62, 65, 70, 74, 77, 81, 97, 103, 104, 109, 111, 113, 122, 127, 130, 140, 146, 166, 175
Root Dipping Technique, 105
root growth, 9, 70, 103, 109, 113, 127, 130
root hair, 109, 122
root system, 49, 77, 104, 166
roots, x, 9, 10, 18, 34, 38, 46, 47, 49, 54, 60, 61, 62, 63, 64, 65, 66, 67, 69, 73, 77, 79, 80, 86, 92, 104, 105, 107, 108, 111, 119, 124, 133, 140, 157, 158, 166, 175
roses, 27, 29, 43, 54

S

safety, 66, 126, 178
saline water, 10, 11, 25
salinity, 10, 11, 15, 23, 27, 29, 52, 54, 93, 126, 133, 163, 170, 181
salinity levels, 11
salt accumulation, 23, 156, 177
salt tolerance, 133
salts, 22, 28, 146
Saudi Arabia, 57
scaling -up, 121
scarcity, 3, 107
science, 87, 127
scope, 3
sea level, 154
seafood, 120, 122
seaweed extracts, 113, 116, 123, 125, 126
Seaweed Waste, 113
security, 102
seed, 112, 113, 127, 164
seedlings, 52, 53, 55, 66, 70, 105, 133, 175

segregation, 139
sensations, 140
sensing, ix, 2, 17, 31
sensor, 17, 19, 20
sensors, 8, 9, 10, 17, 20, 24, 28, 175
shoot, x, 50, 74, 82, 83, 84, 85, 113, 155
shoots, x, 46, 73, 79, 80, 82, 86, 155, 169
showing, ix, 58, 139, 157, 159, 165
signal transduction, 124, 162
silica, 41, 43, 48, 55, 131
Silicon (Si), v, vii, ix, 33, 35, 41, 42, 43, 44, 45, 46, 47, 48, 49, 50, 51, 52, 53, 54, 91, 109, 129, 151
Silicon Sources, 43
SiO_2, 43
small pots, 105
SNP, 66
SOD, vii, ix, 48, 58, 61, 63, 64, 66, 67, 81
sodium, xii, 24, 44, 66, 99, 113, 122, 128, 130, 132, 138, 143, 144, 145
software, 61
soilless cultivation, 12, 18, 22, 92, 109
solid waste, 139
solution, vi, vii, ix, x, xi, xii, 3, 5, 10, 11, 22, 29, 33, 34, 35, 36, 38, 39, 40, 42, 43, 44, 45, 48, 51, 52, 60, 66, 69, 73, 74, 77, 80, 82, 83, 86, 92, 93, 94, 95, 96, 97, 98, 103, 104, 105, 106, 107, 109, 112, 119, 120, 124, 131, 137, 138, 140, 142, 145, 146, 147, 149, 153, 156, 160, 163, 165, 177
solvents, 76
South Africa, 155
South America, 154
Spain, 14, 27
speciation, 79, 80, 83, 86, 134
species, ix, x, 24, 33, 34, 35, 36, 41, 42, 52, 53, 54, 59, 66, 70, 74, 75, 81, 83, 84, 86, 87, 88, 92, 109, 113, 154, 167, 169, 179
species richness, 154
spectra analysis, 85
spectroscopy, 86

Index

spine, 157, 159, 162, 170, 177, 180
spines, 155, 156, 159, 162, 168, 169, 177, 179
stability, 45, 164
Stabilized OSA, 44
standard deviation, 80
state, 17, 75, 80, 112, 134, 140
stock, 93, 96, 104
stomata, 3, 15, 122
Stomatal conductance, 49
storage, 19, 65, 140, 141, 143, 144
stress, viii, ix, xi, 1, 2, 4, 5, 9, 10, 18, 19, 20, 21, 22, 24, 25, 26, 27, 30, 35, 41, 43, 47, 50, 51, 52, 53, 55, 66, 68, 69, 70, 81, 92, 99, 115, 128, 130, 131, 133, 146, 159, 166, 168, 179
stroma, 34
structure, 16, 50, 78, 84, 155
substrate, viii, ix, 1, 2, 5, 6, 8, 9, 10, 11, 12, 16, 22, 23, 25, 38, 40, 78, 120, 133, 146, 151
successive approximations, 6
sugar beet, 109, 132
sugarcane, 93, 95, 96, 146
sulfate, 38, 134
sulfur, 70, 140
superoxide dismutase, vii, ix, 48, 58, 61, 81
supervision, 108
supplementation, xi, 42, 92, 95
susceptibility, 113
sustainability, 134
symptoms, 22, 34, 36, 38, 39, 40, 107
synergistic effect, 65, 70
synthesis, 36, 69, 113, 115, 123, 133

T

tanks, 119, 139
techniques, vii, viii, xiii, 1, 5, 6, 20, 23, 75, 92, 103, 173, 177
temperature, xiii, 3, 10, 13, 16, 17, 18, 25, 29, 31, 39, 82, 106, 144, 156, 157, 160, 165, 168, 169, 173, 174
tissue, 36, 50, 92, 94, 99, 110, 123, 129
tobacco, 129
toxic effect, vii, ix, 33, 48, 80, 87
toxic metals, 65, 75, 88
toxicity, xi, 34, 35, 36, 38, 39, 41, 45, 47, 49, 50, 51, 52, 53, 54, 59, 60, 65, 66, 68, 69, 70, 71, 74, 75, 81, 92, 93, 95, 109, 132, 170
toxicology, 70
trace elements, 65, 74
traits, 70, 96, 127
transformation, 41, 171
transformation processes, 41
transformations, 126
translocation, 65, 70, 79, 80
transmission, 7
transpiration, viii, 2, 3, 4, 5, 6, 7, 8, 9, 11, 13, 15, 16, 18, 23, 25, 26, 27, 28, 30, 31, 38, 45, 49, 71, 77
transport, 53, 81, 133, 146
transport processes, 133
treatment, xi, xii, 20, 60, 61, 66, 67, 74, 86, 94, 97, 98, 102, 104, 112, 120, 122, 134, 138, 140, 142, 144, 149, 150, 157, 161, 165
treatment methods, 74
tryptophan, 111
turgor, 25
Tyrosine, 115, 118

U

U.S. Department of Agriculture (USDA), 134
ultrastructure, 51
United Nations, 98, 134, 141, 149
urban, 103, 119, 120, 121, 123, 126, 131, 133

urban agglomerations, 131
urban areas, 120
urine, vi, vii, xii, 137, 138, 139, 140, 141, 142, 143, 144, 145, 146, 147, 148, 150, 151
Uruguay, 149

V

Valencia, 14, 170
vapor, viii, 2, 6, 13, 15, 20
variables, 4, 14, 146, 166
variations, 19, 25, 26, 109
vegetables, xii, 37, 119, 120, 121, 126, 133, 142, 155, 156, 157, 163, 164, 170, 173, 174, 175, 176, 177, 179, 180
vegetation, 16, 21
ventilation, 3, 6, 17, 31
vessels, 139, 166
vinasse, xi, 92, 93, 95, 96, 97, 98, 99, 132
Vinasse, 93, 96
vitamins, 36

W

waste, vii, xi, 77, 92, 93, 94, 95, 96, 97, 101, 102, 106, 111, 112, 114, 115, 118, 119, 120, 121, 124, 128, 139
waste management, xi, 102, 128
Wastewater, vi, vii, xi, xii, 75, 91, 92, 94, 95, 96, 97, 98, 102, 103, 110, 111, 112, 115, 119, 120, 122, 123, 125, 126, 128, 134, 101, 137, 138, 139, 140, 141, 142, 143, 144, 145, 146, 147, 148, 149, 150, 151

water, viii, ix, x, xii, 1, 2, 3, 4, 5, 6, 8, 9, 10, 11, 13, 15, 16, 17, 18, 19, 20, 21, 22, 23, 24, 25, 26, 27, 28, 30, 32, 42, 43, 44, 45, 49, 51, 53, 61, 66, 73, 74, 76, 77, 79, 82, 83, 86, 88, 91, 92, 93, 94, 96, 97, 98, 102, 103, 104, 105, 106, 107, 108, 110, 119, 120, 124, 132, 133, 134, 139, 141, 146, 147, 153, 156, 157, 159, 166, 176, 177
water quality, 51
water resources, 3, 94, 132
waterways, 110
web, 124
weight changes, 4
weight loss, 9
wind speed, 11
worms, 106

X

xylem, 43, 109

Y

yield, 2, 5, 8, 9, 10, 11, 23, 27, 29, 30, 31, 41, 52, 53, 55, 81, 92, 98, 106, 107, 108, 116, 128, 148, 155, 166, 171, 181

Z

zero-emission, xi, 102, 121
zinc, 44, 52, 69, 70, 87, 109, 132